Extinctions

Are we now entering a mass extinction event? What can mass extinctions in Earth's history tell us about the Anthropocene? What do mass extinction events look like, and how does life on Earth recover from them? The fossil record reveals periods when biodiversity exploded, and short intervals when much of life was wiped out. In comparison with these ancient events, today's biotic crisis has not yet reached the level of extinction to be called a mass extinction. But we are certainly in crisis, and current parallels with ancient mass extinction events are profound and deeply worrying. Humanity's actions are applying the same sorts of pressures – on similar scales – that in the past pushed the Earth System out of equilibrium and triggered mass extinction events. Analysis of the fossil record suggests that we still have some time to avert this disaster: *but we must act now.*

MICHAEL HANNAH is Associate Professor in the School of Geography, Environment and Earth Sciences at Victoria University of Wellington in Aotearoa/New Zealand. He completed his PhD at Adelaide University, specialising in palaeontology and biostratigraphy (the art of dating rocks using fossils). After a brief stint in industry, he took up a position at Victoria University, where he became involved in two major Antarctic drilling projects, helping to decipher ancient changes in climate and the history of the Antarctic ice sheets. Throughout his career he has been fascinated by the story of the evolution of early life and the terrifying consequences of the mass extinctions that are evident in the fossil record.

'... a useful and succinct summary of the research into the reality and timing of mass extinctions from the early concepts to recent research – it brought me up-to-date with current thinking on mass extinctions. I admire his 'sceptical' stance: attempting to discriminate what a mass extinction actually *is* – outside the biggest three – is not as easy as has been assumed. The mass extinctions of the past clearly have relevance to the current approaching catastrophe in the Anthropocene, and the careful appraisal of exactly where we are in comparison with previous extinctions will be of great concern to those interested in the 'long view'. I particularly appreciated the focus on the notion of the interconnectedness of Earth systems.'

– Richard Fortey, author of *Life: An Unauthorised Biography History* and *Trilobite: Eyewitness to Evolution*

'Despite its sombre title and topic, *Extinctions* is an exuberant road trip through the history of life on Earth, led by a friendly and knowledgeable guide who knows all the locals along the way. Visiting so many ancestral Earthlings and vanished ecosystems is heady – and deeply humbling.'

– Marcia Bjornerud, Lawrence University, author of *Timefulness* and *Reading the Rocks*

'Most of life may well be extinct, because of the huge age of the Earth, but Michael Hannah shows vividly in this book that the 8.7 million species on Earth today are profoundly at risk; the lessons of the fossil record tell us what will surely happen if we continue pushing species after species to the brink.'

– Michael Benton, University of Bristol, author of *Dinosaurs Rediscovered*

'Without death, there can be no change. And, as Michael Hannah makes clear in his engaging new book, mass extinctions on various scales have been key shapers of the world as we know it. Had the dinosaurs not abruptly disappeared, we humans would not be here today. But as Hannah also shows, there is something dreadfully menacing about the massive species loss and climate change the world is currently experiencing, making his book a balanced yet deeply unsettling account of what humans are unwittingly doing to the world.'

– Ian Tattersall, American Museum of Natural History, co-author of
The Accidental Homo Sapiens

'An accessible and authoritative guide to the past, present, and future of extinctions. Michael Hannah dives into the fossil record and surveys the great mass extinctions of Earth history, from the death of the dinosaurs to the demise of the woolly mammoth, and explains how they are relevant to understanding the predicament we are in today, and to plotting a better future.

– Steve Brusatte, University of Edinburgh and New York Times/Sunday
Times best-selling author of *The Rise and Fall of the Dinosaurs*

'Michael Hannah's book expertly examines the geological record of mass extinction events. It asks us to consider whether we wish to join asteroid strikes and massive volcanic eruptions as causes of mass extinction. Or whether we can change our relationships with the wonderful diversity of life around us to avoid such an ignominious outcome.'

– Mark Williams, University of Leicester

Extinctions

Living and Dying in the Margin of Error

MICHAEL HANNAH

CAMBRIDGE
UNIVERSITY PRESS

CAMBRIDGE
UNIVERSITY PRESS

University Printing House, Cambridge CB2 8BS, United Kingdom

One Liberty Plaza, 20th Floor, New York, NY 10006, USA

477 Williamstown Road, Port Melbourne, VIC 3207, Australia

314–321, 3rd Floor, Plot 3, Splendor Forum, Jasola District Centre,
New Delhi – 110025, India

103 Penang Road, #05-06/07, Visioncrest Commercial, Singapore 238467

Cambridge University Press is part of the University of Cambridge.

It furthers the University's mission by disseminating knowledge in the pursuit of
education, learning, and research at the highest international levels of excellence.

www.cambridge.org
Information on this title: www.cambridge.org/9781108843539
DOI: 10.1017/9781108919012

© Cambridge University Press 2021

First published 2021

Printed in the United Kingdom by TJ Books Limited, Padstow Cornwall

A catalogue record for this publication is available from the British Library.

ISBN 978-1-108-84353-9 Hardback

Kia whakatōmuri te haere whakamua

I walk backwards into the future with my eyes fixed on my past

MĀORI PROVERB

The longer you can look back, the farther you can look forward

WINSTON CHURCHILL

For June

Contents

Preface

To a first approximation all life on Earth is extinct

David Raup

One estimate suggests that there are 8.7 million species alive on the Earth today.[1] In all honesty, this is our current educated best guess at the total biodiversity of the planet. However, if there are 8.7 million species alive on Earth today, approximately 3.7 billion years ago there was one. That single species evolved and flourished and is the ultimate ancestor to all living things today.

The quote above attributed to David Raup reflects a very palaeontological perspective on life. Raup's somewhat depressing conclusion about the amount of life currently on Earth is based on his estimation that of all the species that have ever existed on the planet, over 99% are extinct. In other words, the 8.7 million descendants of that single first species represent less than 1% of the number of species that have evolved and gone extinct since life first appeared on Earth. If we assume (generously) that all species alive today represent exactly 1% of all the species that have ever lived, simple arithmetic suggests that about 870 million species have lived on the Earth at some point in time. The margin of error on the estimate of 8.7 million species currently alive is ±1.3 million, or about 15%. Applying this level of error to my simplistic estimate for all species that have ever existed suggests a margin of error of about ±130 million species, well above the most outlandish estimates of the planet's current level of biodiversity. From the perspective of deep time, we, and every other organism on the planet, are *living and dying in that margin of error.*

---

[1] This estimate by Camilo Mora and his co-workers (which is not without its critics) is said to be accurate to within ±1.3 million. I chose this estimate because it was (1) fairly recent and (2) well documented.

It's highly likely that most of the millions of species that have gone extinct over the eons of geological time have left no trace of their existence. But some of them – certainly the minority of them – have been preserved in rocks and become part of the fossil record. That record has its limitations, but it has allowed palaeontologists to piece together, often in fine detail, the history of life on Earth. We have recovered fossils of organisms from rocks 3.7 billion years old, some of the earliest living things on the planet. We have fossils that trace the evolution of this early life, and we can watch as it became more complex and diverse, eventually filling the oceans. Through the fossil record we can witness life creeping out of the oceans and onto the land. The record contains fossils of the first flowering plants and a full record of the evolution of whales. In recent years, palaeo-anthropologists have identified an extraordinary array of hominin fossils that, together with genetic data, have triggered a revolution in the understanding of our own evolution. Many of the stunning fossils that have been recovered have captured the public's imagination. It seems as if every child I meet over the age of five knows more about dinosaurs and their extinction than I do.

But aside from the spectacular fossils it contains, the fossil record reveals something more fundamental about the history of life itself. It clearly shows that the journey from the one species living 3.7 billion years ago to the 8.7 million species existing on Earth today has been far from straightforward. In fact, the history of the diversity of life on Earth, as revealed in the fossil record, is a turbulent one. There have been periods when the diversity of life simply exploded, with many new species appearing relatively quickly. But there have also been times when the Earth's biota went through periods of great dying: geological instants when huge numbers of species went extinct simultaneously, causing the planet's biodiversity to tumble.

This book aims to show that this tempestuous history of biodiversity is important. Today, humans are warming the planet: the ice sheets are melting, and sea levels are rising. The chemistry of the oceans is changing – they are becoming more acidic, and areas of low oxygen concentration are spreading. And it's not just that we are changing the climate. To feed and house our ever-expanding population, we are radically changing the way the land is used. Forests are being cut down to make way for agriculture; cities are expanding, reducing the natural habitats that support much of the Earth's biota. As a result, many species have already gone extinct, and many more are under threat. The history revealed in the fossil record offers us a chance to set the current state of the planet into its full historical context. This, in turn, can provide us with lessons that can both help us to understand how we got into our current situation, and offer some guidance about what we might expect in the future, and how we can change it.

Acknowledgements

This book had its genesis many, many years ago when Kim Sterelny – a philosopher of science who seems to be able to produce an erudite book every year – suggested I write a book about mass extinctions. The book you have in front of you does have a lot in it about mass extinctions, but it turned out they were only part of the story I wanted to tell. However, it has taken a long time for me to translate my vague notions of what I wanted to write about into the book you now hold. During that time, I suspect that a lot of my friends, colleagues, and students got heartily sick of hearing me talk about it. They have all been extremely tolerant, and I have to thank them for that. Many of the ideas included here were first tried out on my undergraduate classes. Standing in front of a large class, it was easy to see which ideas were received with enthusiasm and which resulted in almost terminal boredom. If nothing else, the use of cell phones is a great indicator of interest. Interested or not, many of these former students who happen to read the book will probably recognise some of my rubbish jokes (sorry).

Soon after I started writing the book, my confidence faltered, and I almost gave up. Instead, a science writing course offered at Victoria University of Wellington's International Institute of Modern Letters provided a real antidote. I recommend it highly. Course coordinators Ashleigh Young and Rebecca Priestly along with my fellow CREW 352 alumni provided confidence-building encouragement when I needed it the most. They also delivered useful critiques of early drafts of the initial chapters, convincing me that I could write a sustained piece of work that people would want to read. Rebecca Priestly deserves special thanks for introducing me to the amazingly helpful Sam Elworthy at Auckland University Press. Under Sam's careful but firm (very firm when I think about it) guidance, my collection of not very well linked chapters became the backbone of this book.

Four of my colleagues supported me throughout the writing process: James Crampton, Chris Clowes, Ben Hines, and the astonishing Katie Collins. More than any others, these four put up with me rabbiting on endlessly about mass extinctions, the Anthropocene, and other palaeontological subjects, then convinced me that there was a book in there somewhere. James and Chris deserve special acknowledgement. Together we discussed many of the topics covered here, and they helped me to formulate and sharpen my ideas. They also passed along any scientific papers that they thought would be useful for me, and they usually were. James also allowed me to include many of the topics covered in this book in our joint postgraduate course where I had many of my ideas severely challenged – students can do that. All four read early drafts of the book and provided extremely useful feedback. As is usual, any mistakes that are still included here are my fault, not theirs.

There have been lots of useful discussions about many aspects of early life on Earth, mass extinctions, and ancient levels of diversity over beer and wine during meetings of the Palaeontologists in the Pub group. As well as those listed above, stalwarts included Thomas Cooper, Matt Ryan, Joanna Elliot, Lockie Hobbs, Tom Womack, Sonja Bermudez, Shelby Stoneburner, and Lisa McCarthy. I owe you all a beer.

At Cambridge University Press, Matt Lloyd has been enthusiastic about this project from the very beginning and has weathered my appalling proofreading with good humour. As a result of his geological training, he has also been able to pick up several errors of fact and terminology, which is appreciated. Sarah Lambert provided significant help with the nuts and bolts of assembling the book, from producing the index to obtaining copyright permissions. Lindsay Nightingale did an amazing job of editing the manuscript – saving me from several potentially embarrassing glitches and improving the book considerably.

I'd like to acknowledge Craig Hinton-Taylor, head of the IUCN Red List. He not only granted permission for the use of the list's data in Table 1.1, he also provided me with extra material, and picked up a mistake in my calculations and corrected it!

The translation of the Māori whakataukī (proverb) comes from a paper by Lesley Rameka of the University of Waikato. Although the paper concerned early childhood education, it introduced me to the Māori view of time which sees the past, present and future as intertwined. The past informs both the present and the future: one of the key themes of this book.

My whānau have supported me through all the writes and rewrites that the book has gone through. I need to acknowledge my children, Lachlan and Rebecca, and their partners Hannah and Dan. Thank you all so much. And finally, June, my long-suffering spouse – I cannot thank you enough for your continued support and encouragement. Even when I doubted my own ability to finish the thing, you never faltered. I couldn't have done this without you.

Further Reading

Throughout this book, I have resisted the use of formal scientific referencing because I believe that a reader who is not used to the style would find it distracting. Instead, I have listed at the end the scientific sources I found useful in shaping my thinking on the various topics covered in this book. Most are formal scientific publications, but there are also many books – most of which were written for non-scientific readers.

Introduction

We are probably living in the newest period of geological time, the Anthropocene. This interval of time is in the process of being formally established by the geological community to reflect the extraordinary impact that humans have had on the planet. Future geologists will examine rocks of Anthropocene age and, preserved in them, will recognise the massive changes to the planet's ecosystem brought about by human activity. They will find the remains of huge mega-cities and the debris of manufacturing. The sediments will clearly show immense changes in land use, with forests and grasslands abruptly giving way to the intensive agriculture needed to feed an exploding population. Their sophisticated geochemical analyses will show how the planet underwent rapid warming, the result of humans injecting excess carbon dioxide into the atmosphere. Future palaeon-tologists will note a significant shift in the fossil record. In older sediments deposited prior to the Anthropocene, they will recover fossils reflecting a diverse mammal fauna. But that will change as we enter the Anthropocene. Here, the rocks will yield a mammal fauna much less diverse than the one that preceded it. Not only that, but the fauna will be dominated by fossils of domesticated stock – sheep, pigs, and cattle. And this loss of diversity won't be confined to mammals. The record will show a massive loss of biodiversity across all biotic groups in response to deteriorating environmental conditions.

Today we are at the start of that crash in biodiversity. Species are going extinct at an ever-increasing rate. But to appreciate the full impact of human activity on the planet's biota, we need to look beyond simply the number of species going extinct. We must add to it the even larger number of species that are under severe threat of extinction. Biologists have come up with a new word for this combination of extinction and threat – *defaunation*.[1]

But today's defaunation event is not the first time that the planet's biodiversity has crashed; it's not even the first one that humans have been implicated in. It has been suggested that on up to nineteen previous occasions, large parts of the planet's ecosystem have been lost. These ancient events are called mass extinctions and are, in turn, just part of a vast history of biodiversity that begins with the appearance of the first living cell about 3.7 billion years ago.

Our primary source for the history of biodiversity is the fossil record, the remains of past organisms contained in sedimentary rocks. The extraction of this history from the fossil record is difficult. Some aspects of the record are problematic, which limits our understanding: for example, it is biased towards marine animals with hard parts such as shells. Nevertheless, the study of the deep history of biodiversity has consumed the careers of many palaeontologists and other Earth scientists. And, despite the limitations imposed by the record itself, they have been able to document, often in detail, the rollercoaster ride that biodiversity has taken over those millions and millions of years: explosive radiations as evolution explores new ecological niches and the horrendous crashes that result from mass extinctions.

The planet on which life first emerged is a vastly different one from the one we find ourselves on today. At about the time that life first appeared, the Earth's thin primordial crust had started to differentiate. Like curdling milk, rocks containing abundant lighter elements, such as oxygen and silica, separated from rocks dominated by heavier elements, like magnesium and iron. The lighter rocks coagulated together into thick clumps, the cores of what would become today's continents. The rocks made of heavier material formed the crust that

---

[1] It's important to note that despite the use of the word 'fauna' in defaunation, I do include plants (flora); I use it to refer to the planet's entire biota. I tried to invent a new term, one that encompassed all life, but every one I came up with sounded wrong.

makes up the floors of today's oceans. There were oceans on the early Earth as well, separating the nascent continents. They were filled with water, probably transported to the planet largely by comets. Volcanoes spewed forth massive amounts of toxic gases into the thin atmosphere, which was dominated by nitrogen, methane, and ammonia, with little or no free oxygen. On the early Earth, the Sun was less intense, but temperatures still reached up to 70 °C. The planet spun on its axis faster – each day was shorter than today.

Yet this violent, unpleasant landscape was home to our ancestors. They may have been only simple cells, but one of them, just one of them, is the ancestor of all life on Earth. That cell was successful at surviving, so it was able to produce descendants. Some of that first generation in turn were able to give rise to another generation, developing a thread of organisms, each a survival success, linking each generation through time. That first single thread linking one successful organism after another was just the start; given time and evolution, new species appeared, and the thread split. And it split again and again, as the development of the Earth's ecosystem began, and biological diversity rose.

It took a while for life to progress beyond the level of a single cell. But about 600 million years ago, multicellular life evolved. This step-up in complexity from single cells to multicellular animals coincided with an increase in the level of available oxygen. However, as we will see, the history of oxygen is complex. We do know that the tell-tale geochemical signature of the persistent presence of oxygen in the atmosphere is first seen in rocks that are about 2.45 billion years old. However, all the indications are that between 2.45 billion years and 600 million years ago the level of oxygen remained very low. Then, just as the first multicellular organisms appeared, the level of oxygen started to rise. This isn't a coincidence; complex life needs higher levels of oxygen to live. What is surprising is that the evolution of complex life did not meekly follow the steady increase in oxygen. It's more likely that the increasing complexity of life altered the environmental conditions, allowing the oxygen to accumulate.

The first multicellular organisms on Earth were life in slow motion – a sluggish, passive affair, where strange animals were filter feeders firmly attached to the sea floor, grazers slowly eating their way across ubiquitous algal mats, or stationary quilted forms that simply absorbed nutrients directly from the ocean. Then, about 550 million years ago, all hell broke loose, and biodiversity exploded. The oceans rapidly filled with new creatures, and most of the major animal groups suddenly appeared. For over 20 million years, the level of biodiversity increased at a rate that has never been matched since. Life started to move, and we can now recognise animals that crawled across the sea floor and some that burrowed into it. Others left the sea floor behind and swam. Among these were the planet's first predators. For the first time in Earth's history, we can recognise a whole complex ecosystem. It only took about 3 billion years to appear.

But life wasn't the only thing evolving: the planet itself was changing. Its atmosphere, oceans, and rocks were being transformed, often in response to the appearance of ever more complex organisms. This transformation of the planet is part of the development of the complex mechanism that is responsible for maintaining environmental conditions on Earth. More importantly for us, it maintains conditions that are suitable for the continuation of life on Earth. This automatic climate control mechanism is called the Earth System, and life has not been a passive player in its development. The evolution of increasingly complex organisms has fundamentally altered the physical make-up of the planet. The story of oxygen, which we will discuss later, is a great example of the link between physical processes and life. The Earth System existed in a rudimentary form for the first 3 billion years of the planet's history. But it wasn't until about 550 million years ago, at the same time as the burst of diversity and the appearance of the first complex ecosystems, that the Earth System became fully operational. Since then, the system has, sometimes against all the odds, and often not perfectly, maintained a relatively constant set of environmental conditions on Earth.

What caused that explosion of biodiversity 550 million years ago that ushered in a fully functioning Earth System remains unclear. It's possible that the level of oxygen reached some critical threshold, or there was a sudden increase in the supply of nutrients to the ocean. It could even be something as simple as the development of hard parts, which opened new evolutionary opportunities and made the preservation of these animals as fossils so much easier. Whatever it was, it came with a bonus – a much-improved fossil record. The biases inherent in the record didn't go away, but the record became more complete. By this stage, there were enough fossils being preserved to allow palaeontologists to carry out increasingly sophisticated estimates of the changes in biodiversity through time. Assisting in this analysis has been the development of large databases that track the evolutionary appearance and extinction of organisms. Palaeontologists are now able to apply statistical analyses to the record. This has allowed the rise and fall of biodiversity over the past 550 million years or so to be estimated in some detail.

The results of these analyses confirmed the presence of mass extinctions – sharp drops in biodiversity, often followed by almost equally rapid rises. The first real attempt to document these events in the 1980s suggested that there had been five such events over the past 500 million years – now forever popularised as the 'Big Five'. The largest of these occurred at the end of the geological period called the Permian (about 250 million years ago). The end-Permian mass extinction resulted in the loss of up to 96% of marine species (with correspondingly high levels of extinctions on the land). But the most famous of the Big Five is undoubtedly the mass extinction that occurred at the end of the Cretaceous period some 66 million years ago. It had a lower kill rate; current estimates of species that went extinct range between 65% and 75%. But it included the charismatic dinosaurs, ensuring its position as the most famous mass extinction of all time.

A mass extinction is defined as a rapid increase in the rate of extinction that occurs on a global scale and involves more than one animal group. Although recovery from these events can be slow, the actual extinctions are always very rapid, with some of the biggest thought to occur over an interval as short as 20 thousand years. Using this definition, one estimate suggests that there have been up to nineteen mass extinctions over the past 550 million years. Although none exceed the 96% kill rate of the end-Permian event, some now seem to be bigger than the other four members of the Big Five group.

Exactly what causes a mass extinction is a fraught subject. Over the years there have been many suggestions; some sensible, others just plain impossible. What we can say is that most mass extinctions of the past 500 million years occurred at exactly the same time as huge volcanic events. These ancient volcanic eruptions resulted in an outpouring of lava on a continental scale and the injection of incredible amounts of carbon dioxide (and other noxious gases) into the atmosphere. Carbon dioxide is a greenhouse gas, and as its level in the atmosphere rises, so do global temperatures. The amount of carbon dioxide released during these events was so large that the temperature rose to a point where the oceans themselves became significantly warmer. A warm ocean cannot carry as much oxygen as a cool one, and this results in widespread anoxia (very low levels of oxygen). As if that wasn't enough environmental stress, the high levels of atmospheric carbon dioxide also caused the acidification of the oceans. Between the acidic, oxygen-poor ocean and spiralling global temperatures, it's safe to say that the global environment would be severely stressed, undoubtedly leading to some extinctions.

In some cases, this highly stressed environment may not have been enough to cause a mass extinction; something else was needed to push the ecosystem over the edge. There are a variety of suggestions for what could provide that final push. In the case of the end-Permian event, the arrangement of the continents appears to have amplified the climatic effects. For the end-Cretaceous mass extinction, the impact of a meteorite some 10 kilometres in diameter delivered the coup de grâce.

So, what follows an event that has resulted in the death of up to 96% of life on Earth? There are many images in the literature that attempt to illustrate the devastation caused by this mass extinction. One image shows a few fish swimming over an almost empty sea floor where once coral reefs flourished. In another, rare reptilian survivors stagger across a broken landscape of rotting vegetation that was once a forest. As overly dramatic as these pictures sound, they are probably not far from the truth. But no matter how bad the devastation resulting from a mass extinction, life always bounced back. However horrendous we think a mass extinction is, as far as the history of life is concerned, it is just a temporary setback.

But that's life in the broadest sense. When examined in detail, it is clear that a mass extinction can completely change the make-up of the planet's biota. Basically, the assemblage of organisms that evolves after a mass extinction is never the same as the one that was decimated by it. Each mass extinction results in a wholesale clearing of ecological niches. Following the extinction event, these empty niches are available to be filled through the evolution of new, unrelated species. However, we can't pick winners; there is no way to decide before a mass extinction which species will be killed off and which will survive and diversify.

There is one mass extinction event that deserves to be singled out. It may not be as large as the other events we regard as mass extinctions, and it didn't involve dinosaurs, but it was significant. What makes this event so special is that humans are implicated in causing it. Prior to 50 thousand years ago, giants roamed the world. In Australia, kangaroos 3 metres tall lived alongside 7-metre-long monitor lizards and wombats the size of hippos. North America was inhabited by elephant-like mastodons and mammoths, which were preyed on by sabre-tooth cats and dire wolves.[2] Rhinos, cave bears, and giant Irish elks roamed Europe. Fifty thousand years ago, these animals, collectively known as a megafauna, started to die out. By 11 thousand years ago, most of these giant mammals were extinct.

---

[2] Yes, there really were dire wolves.

The cause of the megafaunal extinctions is still under intense discussion. Climate change certainly played a part. Over the past 50 thousand years, the Earth's climate has shifted dramatically. For most of that time, the planet was in the grip of an ice age, with glacial ice covering a great part of the globe, reaching a maximum extent about 20 thousand years ago. Following the last glacial maximum, climate generally warmed. However, superimposed on this warming trend were some significant fluctuations as the planet's climate bounced wildly from periods of intense cold and intervals of warming. Climate stabilised about 11 thousand years ago, about the time that the megafaunal extinctions ended. These rapid changes in climate must have placed severe ecological pressure on the megafauna. But, on their own, the changes in climate are unlikely to have been enough to trigger the megafaunal extinction event. After all, the megafauna had survived through several previous ice ages. The extra element that finally precipitated extinctions appears to be the arrival of humans, *Homo sapiens*. The time interval we are interested in coincides with the main migration of our direct ancestors out of Africa. It seems likely that the combination of fluctuating climate and the arrival of humans triggered the biggest extinction event in 50 million years.

Some scientists have argued that the megafaunal extinction represents the beginning of the Anthropocene defaunation. And they have a point. Sedimentary rocks of the past 11 thousand years do record the increasing influence of human activity on the environment. The spread of agriculture, the domestication of animals, and the development of industry and manufacturing have all left indelible markers in the record. But the environmental damage caused by these ancient humans pales in comparison to the damage being wrought today. I think of the last 11 thousand years as the slow fuse that leads to the explosion of human activity that defines the Anthropocene.

There are aspects of today's Anthropocene defaunation that are similar to the ancient mass extinctions. Firstly, the rate of extinction has rapidly increased. One estimate suggests that the rate at which amphibians are going extinct today is about 100 times faster than the pre-human rate. Secondly, the extinctions are not restricted to certain areas; they are occurring on a global scale. If it is a mass extinction, it's the youngest in the history of biodiversity. This has led to it being referred to as *the Sixth Extinction*, a nod to the original Big Five. However, the percentage of extinct species recorded today is far below the 65% to 96% level of the big ancient mass extinctions. Perhaps we can take comfort in that. But what is happening now is defaunation, and this means we shouldn't restrict our attention to the extinctions. By accepting the concept of defaunation, we must expand our focus to include the many species that are undergoing huge reductions in population size, placing them under severe threat of extinction. If we don't do something to mitigate the situation, we will lose them, and then we will be facing a mass extinction right up there alongside the one that killed the dinosaurs.

Some claim that since life has survived all the ancient mass extinctions, it will survive the current crisis. In their view, we shouldn't worry about our level of extinction growing to match the past mass extinctions; we can manage our way through. On the surface this is true; life has always survived. However, this is a simplistic reading of history. It is true that following a mass extinction event, no matter how large, biodiversity has always recovered. But that's not the whole story. Mass extinctions, especially the big ones, change the composition of the biota. Indeed, it has been argued that most evolutionary innovations are the result of the burst of activity that occurs as the biota recovers following a mass extinction. This means that the make-up of the biota that goes into a mass extinction will be different from the one that comes out the other side. In our case, there is no certainty that any future biota will include humans. We could try to ensure that humans and the species we need for our survival make it through and let the rest die out. This is a horrendous idea because we need a fully functional diverse ecosystem for the Earth System to operate properly. The biosphere is an integral part of the planet's life support system. In addition, no matter how I try to imagine the mechanics of selectively saving only some species, I come to the conclusion that it couldn't be accomplished without the loss of a significant proportion of the human population and an extremely depleted biota. I'm sure there are apocalyptic science fiction novels written about this sort of scenario. Surely it is far better to do something before we get to that point.

One of the most significant differences between the Anthropocene defaunation and the ancient mass extinctions lies in its cause. In one sense, all the ancient mass extinctions are the result of a collapse of at least a large part of the global ecosystem. Setting aside the megafaunal extinction for a moment, all of the ancient mass extinctions were triggered by natural events – massive volcanic eruptions, the arrangement of continents, sea level changes, ice ages, or meteorite impacts – operating either singly or in combination. The megafaunal extinction was in all likelihood caused by a combination of climate change and the appearance of *Homo sapiens*. The Anthropocene defaunation has only one driver: the changes that humans are inflicting on the planet's ecosystem.

At the heart of this book, however, is the message that our situation is not hopeless. It's clear that, compared with the mass extinctions contained in the fossil record, we are only at the beginning of this catastrophe. We have a little time to act before we reach the level of the ancient mass extinctions. We had better use what time we have wisely.

I The Anthropocene and the Earth System

There are no places left on Earth that have not been affected in some way by human activity. Vast areas of the planet are being cleared to provide space both for our ever-expanding cities and for the intensive agriculture needed to support our burgeoning population. Even the supposedly pristine environment of Antarctica has been affected. Ice cores taken from the continent record human events such as the industrial revolution and the dawning of the atomic age. The composition of the very air that we breathe has been altered, largely through the addition of greenhouse gases, which are causing the planet to warm and sea levels to rise. The damage being done to the Earth's environment is profound, and it's leading to an alarming loss of the planet's biodiversity.

In fact, humans started altering our planet's environment thousands of years ago. The record of human activity is a key feature of what is currently the youngest formally recognised interval of geological time, the Holocene, which is considered to have begun about 11,700 years ago (the full Geological Time Scale can be found in Appendix 1). Starting at, or near, its base, sediments of Holocene age record increasing evidence of human activity. In parallel with the increasing human population, simple villages become cities, and small farms give way to intensive agriculture. Records of changing land use, mining, and evidence of human manufacturing – objects like ceramics and glass – all become increasingly common as we approach the present. Traces of human activity start out in localised areas and at a low level, but as we approach more recent times, they become more widespread and intense, reaching a peak in the present day.

In the early 2000s, an atmospheric chemist, Paul J. Crutzen, suggested that the change from the limited record of human influence over the past 11 thousand years to today's massive alteration of the environment implied that we have left the Holocene and entered a new geological epoch that he called the Anthropocene.[1] Both the idea and the name stuck, and in 2009, the Subcommission on Quaternary Stratigraphy (part of the international organisation that oversees changes to the Geological Time Scale) established a working group to formally decide whether the Earth had moved out of the Holocene into a new geological epoch. If they decided that it had, they were to suggest where the base of this new time unit could be placed.

The working group presented its report in 2016. They agreed with Crutzen that the shift in intensity from the low levels of human activity recorded earlier in the Holocene to the level we see all around us today warranted the establishment of a new geological epoch. The Anthropocene had passed its first official hurdle. However, there is continued discussion about where the base of this new time interval should be placed. This is a new *geological* time unit, so the base needs to be identified using criteria that future geologists will be able to see in the rock record. We are looking for a marker that is easily recognised, unambiguous, global in scope, and reflects an event that occurs everywhere on the planet at the same time. There was a strong push to put the base at the beginning of the industrial revolution. However, that varies from place to place. It began in England during the eighteenth century, then spread through Europe and into North America during the 1800s. Some parts of the globe didn't become industrialised until the mid to late twentieth century – and even today, there are areas that are not fully industrialised. A boundary like this, which has different ages depending on where you are geographically, is called

---

[1] The name comes from the Greek *anthropus* meaning human, and -cene from the Greek *kainos*, translated as recent. The use of the -cene ending to an interval of geological time generally indicates that the interval is at the 'epoch' level in the time scale hierarchy.

diachronous, and it is not appropriate for a unit of the Geological Time Scale.

Another suggestion was to place the boundary at the exact moment that the Trinity atom bomb test in New Mexico was detonated. This would mean that the Anthropocene started on 16 July 1945 at precisely 11:29:21 Greenwich Mean Time. This was a good contender as a base because after this event, traces of certain radioactive isotopes that are only produced by the detonation of an atomic bomb start being recorded in sediments. And it certainly was very precise. But there is a problem; for a few years following the test, levels of these isotopes remained at a very low level and are hard to measure. However, their levels started to rise and become much more obvious in the early 1950s. This is also the time, following World War Two, when the level of human activity increased dramatically everywhere, a period referred to as the 'great acceleration'. The upshot of this is that the date that will be chosen for the beginning of the new epoch is likely to be sometime around 1950.

When, as seems likely, the Anthropocene is finally ratified as a geological epoch and formally included in the Geological Time Scale, who will use it? There are a significant number of geologists who don't like the idea at all – they either just don't see the point or object to defining an interval of geological time in advance of it happening – but I suspect that these naysayers are in the minority. As a practising palaeontologist who usually examines fossils from much older sediments, I'll probably have no need for it in my day-to-day work. Other Earth scientists, particularly those specialising in Quaternary studies, who examine lake sediments and ice cores to understand climates of the very recent past, almost certainly will. However, ultimately it doesn't matter whether or not geologists use the new epoch, because it transcends what a period of geological time is all about. It isn't just an interval of time; it is an important signal to humanity. It makes it abundantly clear that we are now living in a world of our own making, and that it is up to us to undo the environmental damage we are causing.

THE EARTH SYSTEM

The increased levels of human activity that are being used to define the Anthropocene are reflected in the serious damage being made to the planet's entire ecosystem. Thirty years ago, scientists attempting to find strategies to mitigate environmental problems would probably have studied individual parts of the planet's ecosystem separately. Physical geographers would look at the land changes, physicists the atmosphere, biologists the biota, and so on. However, so broad are the effects of human activity in recent decades that it was realised that looking at all the different parts of the planet in isolation wouldn't get us very far, and that a more holistic approach was needed. This has led to the application of what is known as *Earth System science*.

The Earth System recognises that the planet is made up of a number of reservoirs or spheres, which are linked to each other via complex feedback loops or fluxes that cycle energy, water, and various elements (including those necessary for life, such as carbon, phosphorus, and nitrogen) through the reservoirs. These fluxes ensure that a change made in one reservoir will flow through and affect the entire system. I've tried to illustrate a very simple model of the Earth System in Figure 1.1. It shows the system as comprising four reservoirs:

- The atmosphere – the sphere of gases that surrounds the planet;
- The hydrosphere – all the water, liquid and frozen, that sits at or near the Earth's surface;
- The biosphere – the Earth's thin coating of living organisms;
- The geosphere – all the rocks and minerals that make up the solid Earth.

There are more complex models of the Earth System where each reservoir is subdivided – for example, the geosphere could be split into the crust, mantle, and core, and the cryosphere (ice sheets, glaciers, etc.) could be split from the hydrosphere, and the relationships between these sub-spheres examined – but we will stick for now with the minimal version. The linking feedback loops are shown by the arrows, and again, in more complex models there are many more of

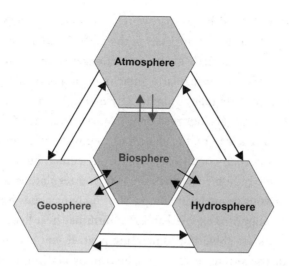

FIGURE 1.1 A very simple model of the Earth System. All the reservoirs are labelled. The arrows that join them represent some of the complex feedback loops (fluxes) that operate to maintain constant environmental conditions on Earth.

them. Although composed of separate parts, for at least the past 600 million years the Earth System has operated as a single integrated whole to keep environmental conditions on Earth relatively constant, and these conditions ensure that the planet is able to support life.

The Earth System approach developed out of James Lovelock's Gaia Hypothesis. In the mid-1960s, Lovelock suggested that we should consider the planet as a single homeostatic system that worked to maintain conditions on the planet that are suitable for life. He likened the planet to a single giant superorganism. Unfortunately, some people took this analogy literally and thought that Lovelock was claiming that the Earth was actually alive (which I don't think he was). And perhaps Lovelock should have stuck with his original name – the Earth Feedback hypothesis. Instead, he named it after a primordial Greek goddess who personified the Earth. A living planet named after the Earth Mother was never going to fly well in the scientific community, and the whole hypothesis quickly became lost in a welter of New Age waffle. But to reject Gaia completely is to throw the baby out with the bathwater: there is no doubt that Lovelock's hypothesis forms the basis of the Earth System approach.[2]

The Earth is not a single giant sentient superorganism. Instead, it's best thought of as a self-regulating machine composed of separate but interconnected parts that automatically maintain environmental conditions. However, Lovelock was right in suggesting that the Earth System does this in a way that is analogous to how a living organism maintains its internal conditions, a process called homeostasis. A simple example of homeostasis is the way you regulate your body temperature. If your temperature drops, your body responds by reducing your blood supply to exposed areas to minimise heat loss. At the same time, you start shivering, generating heat. If you get too hot, your body will try to reduce your temperature by starting to sweat. These are automatic responses to a change in your internal environment – you have no control over them. You don't consciously decide to turn on your sweat glands when you are too hot, but neither can you choose not to turn them on. The Earth System uses an automatic homeostatic-like process to maintain a set of environmental conditions that are suitable for life to exist. It's important to remember that in the Earth System, the biosphere is not simply a passive player responding to changes in the physical environment. The biosphere actively can and does participate in the alteration of the physical world in order to maintain habitable conditions on Earth.

The Earth System didn't just suddenly appear as a fully functional mechanism. As we will see, it emerged gradually, in parallel to the evolution of life on Earth. The fossil record contains spectacular examples of how the evolution of life has driven profound physical

---

[2] The New Age version of Gaia as a loving Earth Mother caring for the planet has infiltrated the mainstream media. I am writing this, at home, during a lockdown caused by the Covid-19 pandemic. I have heard media commentators seriously asking if the pandemic is Gaia's vengeance for the mess we are making of the planet. No, it isn't. For that to happen, the Earth really would need to be a living, thinking organism that instructed a virus found in bats to mutate so it could transfer to pangolins, then ask it to mutate again in order to infect humans. But the planet isn't alive, and no animal plans its evolution.

changes to the environment. If evolution makes it clear that all living organisms are connected, descending from a single common ancestor, the Earth System takes this further. It demonstrates that life is intimately connected to the physical world. A complex, fully functional ecosystem has a central role to play in maintaining the equilibrium of the Earth System. To emphasise this, I have put the biosphere at the very heart of my simplified model in Figure 1.1.

The Earth System has, despite purely physical changes such as a steadily brightening Sun and tectonic upheavals, maintained planetary conditions in a state of equilibrium. This means that if it is temporarily perturbed by some sort of physical agency, such as a massive volcanic eruption that alters the environmental conditions, the system will use the automatic homeostatic processes to reverse the change and restore the original conditions. Think of it as a marble sitting at the bottom of a shallow bowl. It is in equilibrium: as long as nothing changes, it will stay right where it is. However, if you push the marble up the side, it is no longer in equilibrium. When released, the marble will try to return to equilibrium by running back down to the bottom of the bowl – in fact, it will overshoot and run back and forwards, but eventually it will settle at the base again.

An equilibrium state can change over time. A long-term perturbation – more properly called a forcing – may change the conditions to such a degree that some tipping point or threshold is exceeded, and the system cannot quickly restore the original conditions. If we go back to our marble example, it's equivalent to pushing the marble so hard that it tips over the edge and settles down in a new position – a new equilibrium state (it will probably be the floor).

In Figure 1.2, I'm showing a model of an imaginary planet's climate wobbling around one equilibrium point in response to perturbations, then shifting to another equilibrium state altogether in response to a forcing. The dashed line in the figure tracks the changes in a planet's environmental conditions through time. It starts on the left-hand side in equilibrium one. As time passes, small perturbations to the environment push the system out of equilibrium, but it quickly

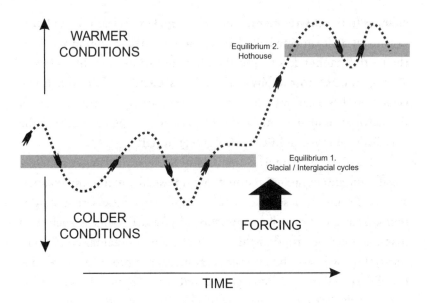

WARMER
CONDITIONS

Equilibrium 2.
Hothouse

Equilibrium 1.
Glacial / Interglacial cycles

COLDER
CONDITIONS

FORCING

TIME

FIGURE 1.2 A cartoon of a shift in equilibrium in the planetary system. See text for explanation.

returns to its original state. I have shown the system as cycling through periods of cold and warmth; you can think of them as alternating glacial and interglacial conditions, although in my example for most of the time the planet is warm with occasional periods of cold. Note how the conditions don't always return to exactly the same state; they wobble around an equilibrium point. The large black arrow is a forcing, some significant environmental shock that pushes the system so far out of its initial equilibrium that it passes some sort of threshold or tipping point and cannot regain its original equilibrium level. In this case, a new equilibrium is established, and, in my example, the new equilibrium results in a more or less permanent hothouse set of environmental conditions. An equilibrium shift like the one shown in Figure 1.2 would cause significant changes in the planet's environment, which would be accompanied by damage to its ecosystem. In the fossil record, these shifts in equilibrium are usually accompanied by a mass extinction. If the forcing were to be removed, given enough time, the system would probably return to something like the original equilibrium point, resulting in a reduction of the planet's temperature – but that could take millions of years.

THE ANTHROPOCENE AND THE EARTH SYSTEM

I can make this more concrete by looking at how the Earth System is operating in the Anthropocene, using the carbon cycle as an example. This important biogeochemical cycle moves carbon through all the spheres that make up our simple model of the Earth System and is fundamental to maintaining life on Earth. It is also one we are currently altering significantly through our burning of fossil fuels. Figure 1.3 shows this cycle, albeit in a simple form. The black lines linking the various spheres represent the fluxes that move the carbon – either as carbon dioxide gas or as part of some other molecule – around the system. I have not included any human-induced alterations to the cycle.

Starting with the geosphere, in the absence of humans, the major source of carbon dioxide on Earth is from gases venting from

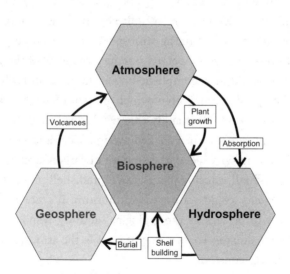

FIGURE 1.3 A simplified illustration of the relationship between the carbon cycle and the Earth System.

volcanoes. The carbon dioxide released from volcanoes enters the atmosphere, and there it plays a significant role in regulating the planet's temperature. It's a greenhouse gas, and the higher its concentration, the higher the temperature of the atmosphere. But the levels of carbon dioxide in the atmosphere don't necessarily continue to grow as volcanoes release additional gas to the system. Plants remove the gas by absorbing it as they grow, and the carbon then enters the biosphere. Atmospheric carbon dioxide is also absorbed into the waters of the world's oceans. But, as is the case with the atmosphere, the level of carbon doesn't just continue to build in the oceans; many organisms strip the carbon from the water and use it to build their shells with calcium carbonate. When plants die, the carbon moves into the geosphere as carbon-rich rocks such as coal. Similarly, when the organisms with carbonate shells die, they accumulate on the ocean floor and can be incorporated into the geosphere as limestone – locking all the carbon away.

Under pre-human conditions, the carbon cycle was kept in balance, maintaining a level of carbon dioxide in the atmosphere that ensured that the temperature on Earth stayed within a range that was suitable for life. If a sudden rise in volcanic activity caused an increase in the level of carbon dioxide in the atmosphere, the machinery of the Earth System would begin to operate to reduce the level of the greenhouse gas. This would be accomplished by an increase in both plant growth and shell production, both of which are stimulated by a rise in the carbon dioxide level. Ultimately this would lead to atmospheric carbon being removed and then locked away as coal and limestone. Once the volcanic activity faded, the amount of carbon dioxide in the atmosphere would fall, causing plant growth and shell production to slow, and the cycle would return to equilibrium. It's not a perfect system. There will be lags as the biological activity responds to an increase in carbon dioxide, so its concentration in the atmosphere will never be absolutely constant.

There is one other process associated with the Earth System that, under certain conditions, can strongly affect the amount of carbon dioxide in the atmosphere. It's not strictly part of the carbon cycle, so I haven't shown it in Figure 1.3. But chemical weathering of silicate rocks (rocks that contain high levels of silica) can have a profound effect on the level of carbon dioxide in the atmosphere. Chemical weathering is a result of carbon dioxide mixing with rainwater to produce a mild acid that attacks the surface of the rocks. As part of the chemical reaction, carbon dioxide is incorporated into the weathering products, which are then washed out into the ocean. This form of weathering can be very effective at drawing down the level of carbon dioxide in the atmosphere, but the conditions have to be right. The appropriate rocks must be available at the surface of the planet, and the climate has to be warm and wet. It also helps if the continents are in the right position on the globe. Nevertheless, as we will see, there have been occasions when chemical weathering has played an important role in the history of life.

I said earlier that Figure 1.2 showed the track of an imaginary planet. But it's actually not too far from what is happening on Earth. Starting about 2.7 million years ago, the Earth's climate has cycled through a series of glacial and interglacial periods, much as I showed in Figure 1.2. Massive ice sheets built up during the cold glacial periods and melted away during milder interglacials. The last glacial period ended at the start of the Holocene, about 11,000 years ago, and we are currently living in an interglacial. If conditions had continued as they had for the past 2.6 million years, we might expect that the Earth would cool and slip into another glacial period. But humans have changed all that.

Gathering pace during the Holocene and exploding into the Anthropocene, human perturbation of the system has been growing, and it's approaching the level of a forcing with the potential of pushing the system through a tipping point. Land clearance, habitat destruction, and forced extinctions are damaging the biosphere, lowering diversity. By burning fossil fuels, we are releasing the ancient carbon that was absorbed by plants and locked away in the geosphere. This has resulted in a rapid increase in the level of carbon dioxide in the atmosphere and, as a result, the globe is warming. The feedback loops, which under normal conditions would bring the system back into equilibrium, are struggling. Plants cannot use the carbon dioxide fast enough. We are compounding the problem by damaging the biosphere through land clearances. This results in a build-up of carbon dioxide in the atmosphere, raising the planet's temperature. Our oceans are absorbing huge amounts of the gas, but as a result their waters are becoming more anoxic and acidic, limiting the ability of shell builders to remove the excess carbon. If we don't take urgent action to reduce the human-induced forcing and allow the biosphere to heal, we may push the system past a threshold from which there is no going back. If we were to allow an equilibrium shift to occur, it would bring with it an entirely new set of very much warmer conditions and – as we will see later – increase the possibility of a mass extinction.

THE ANTHROPOCENE DEFAUNATION

So, how much damage are we doing to our biosphere? Some estimates of Anthropocene extinctions suggest that we are losing between 11,000 and 58,000 species annually. Biologists researching these extinctions have established that the primary drivers are habitat reduction through changing land use, and the introduction of invasive species. However, as well as causing extinctions, these twin drivers can cause a reduction in the population size of those species that do survive, putting them under severe threat of extinction. Biologists have coined a new word for what is happening to our biota during the Anthropocene – defaunation. This term is broader than just the extinctions on their own. It includes all the extinctions together with the many species that are under severe threat of extinction.

The estimate of an annual species loss of between 11,000 and 58,000 is extremely broad. Can't we be more precise? Unfortunately, that's not possible. It is very hard to pin down exactly how bad the Anthropocene defaunation is. This is a result of our astonishing lack of knowledge about the Earth's current biodiversity. We don't even know how many species live on our planet! In this book, I am using an estimate of 8.7 million species alive today. It is, however, only one estimate among many. While most reasonable estimates range from 2 million to 10 million, some outliers get as high as 100 million. This fundamental gap in our knowledge of the planet's biodiversity is compounded by the fact that only a small fraction of living species – only about 20%, by some estimates – has been formally described. Because we need to know that a species exists before we can decide whether it's extinct or threatened, our evaluation of the consequences of the Anthropocene defaunation has to be based on that 20%. One thing we can say, with a fair degree of certainty, is that many species have either gone extinct or are under severe threat of extinction before we had a chance to describe them. We just don't know how many species are involved – but it does mean that any estimate we make of the scale of Anthropocene defaunation will be an underestimate, probably a large one.

The situation isn't entirely hopeless; there are some data available. The Red List[3] is an attempt to assess the level of the threat of extinction of every species on the planet. Again, we must keep in mind that the conservationists who assess the data for the Red List are limited to published species, and this is usually only a very small percentage of known species. In Table 1.1, I have assembled extinction/threat data on several large groups of animals. The first column lists the number of species that have been assessed for the animal groups, with the approximate proportion of species that have been assessed shown as a percentage. In the case of mammals, birds, amphibians, and cephalopods (a group that includes octopuses, squids, and cuttlefish), just about all of the recognised species have been assessed. In the reptiles, well over half have been assessed, but for the remaining groups the data are poor to very poor. The state of

---

[3] The Red List is an amazing resource. Managed by the International Union for Conservation of Nature, it can be found at http://www.iucnredlist.org/

Table 1.1 *The Anthropocene defaunation*

See text for details. Data from the IUCN Red List of Threatened Species (IUCN 2020).

Taxon	Number of species assessed (% of described species assessed)	Extinct and extinct in the wild (%)	Extinct, extinct in the wild, and threatened (%)
Mammals	5,899 (91%)	1.46	23.48
Birds	11,147 (100%)	1.47	14.80
Reptiles	7,833 (70%)	0.42	18.37
Amphibians	6,892 (84%)	0.54	33.56
Insects	9,793 (1%)	0.65	19.23
Bivalves	801 (6%)	4.00	29.09
Gastropods	7,221 (9%)	3.89	32.54
Cephalopods	750 (90%)	0	0.67
Corals	864 (40%)	0	26.85

threat for the insects is particularly badly understood, with less than 1% of species assessed.

During the rigorous assessment process used by the Red List, species are assigned to one of a number of categories depending on the level of threat. Categories range from extinct through to under no threat at all. The categories we are interested in here are: *extinct* – species that we are absolutely confident no longer exist; *extinct in the wild* – species that only exist in zoos, reserves etc; and *threatened*, which includes the subcategories of critically endangered, endangered, and vulnerable. The second column in Table 1.1 shows the percentage of known species that are either extinct or extinct in the wild. The final column adds the threatened category to the extinct and extinct in the wild column – making it a measure of defaunation.

So, based on the Red List data, how bad is the Anthropocene defaunation? Looking first at the extinct and extinct in the wild column, and even given that the values are almost certainly under-estimates, the current situation doesn't look too dire. Estimates of extinction range from 0% (corals and cephalopods) to about 4% (bivalves and gastropods). But to understand our situation fully, we need to look at the whole picture and include the threatened species. Then, with the exception of the cephalopods (for which less than 1% of species are under threat), the situation gets a lot worse. In the other animal groups listed, combined estimates of extinct and threatened species range from 14.8% (birds) up to a whopping 33.56% (amphibians). I find it appalling that nearly 20% of insect species are under threat when only about 1% of known species have been assessed.

Should we worry about this level of biodiversity loss? Some people believe that we needn't. In an opinion piece, entitled 'We don't need to save endangered species. Extinction is part of Evolution' (with a subtitle of 'The only creatures we should go out of our way to protect are *Homo sapiens*') published by the Washington Post in 2017, biologist R. Alexander Pyron suggested just that. He argued that we shouldn't worry about the current wave of extinctions; there have been other mass extinctions in the past, and life made it through. He suggested that the clearing of the ecological deck by a mass extinction could be considered a good thing, allowing evolution to experiment with new forms that would eventually add to biodiversity. Critics wasted no time in telling Pyron that he was wrong, pointing out how important a diverse biota was to the planet.[4] In fact, biodiversity has been described as a central component of the Earth's life support system, underpinning the Earth's entire ecosystem. Our very existence depends on a healthy level of biodiversity. Here's why.

---

[4] Pyron himself later admitted that he was wrong, and in a post on his lab page, he explains that he did sensationalise his argument, and because it was only a short piece, it came out all wrong. He also points out that he didn't choose the title of the article. In fact, he sets out his strong support for the broad conservation of biodiversity. It's called 'Statement on Biodiversity Conservation', and you can find it at http://www.colubroid.org/

While we rely on only a limited number of domesticated plant and animal species for our food supply, they in turn depend upon a host of wild species. Plants, fungi, and bacteria help maintain the soil's fertility, and many species of insects ensure pollination. In Europe, a decline in the numbers of insect pollinators has been directly linked with a significant reduction in the abundance of some plant species. The loss of some small vertebrate species has resulted in whole ecosystems being damaged. We rely on a diverse biota to maintain the quality of water in rivers and lakes, and it can also help strip pollutants out of the atmosphere. Human health is directly affected, as new medicines are very often derived from a diverse array of plant and fungi species. Future medicines might rely on species that we haven't even found or described yet, plants and animals that in the meantime might go extinct.

Perhaps more importantly, a fully functioning ecosystem is a key part of the Earth System. Many of the feedback loops that we rely on to maintain the Earth's environmental conditions cycle elements such as carbon, nitrogen, and phosphorus, all of which are vital for life, through the biosphere. For the biosphere's key role in the Earth System to be effectively carried out, we need a healthy biosphere supporting a diverse ecosystem. The Anthropocene defaunation is as much a destabilising forcing on the Earth System as the addition of greenhouse gases to the atmosphere.

Beyond the direct physical benefits that biodiversity provides for humanity, there is the simple unadulterated joy of walking through tropical rainforests, diving on a coral reef, or walking along the bank of a stream and seeing a beautiful diverse ecosystem in action. It's good for the soul.

HISTORY IS IMPORTANT

The Anthropocene defaunation doesn't exist in isolation; it is part of a longer history of biodiversity that started some 3.7 billion years ago. Today we have the tools to extract this history from the fossil record and lay it out in some detail. If we want to avoid or mitigate the worst of the environmental problems facing us, then this history is important. It can shed light on our somewhat precarious position and offer some indication of what the future holds for life on Earth. Through it, we can examine the response of ancient biotas to significant shifts in the Earth System. We can look in detail at ancient mass extinctions. Do they look similar to the Anthropocene defaunation? Were these extinction events restricted to one specific area or did they affect the entire globe? How does the biosphere recover from a mass extinction? The understanding we can gain through a detailed look at the fossil record can help us plan for what may happen in the future. We are forcing change onto the Earth System, and the fossil record contains information that shows us just how resilient the Earth System is to change – how far can it be pushed without it shifting to a new equilibrium state, triggering a mass extinction?

However, there are limitations to documenting and understanding this history, most of which are due to the vagaries of our source material, the fossil record. If we are to understand what these limitations are and why the record imposes them, we need to take a short detour and spend some time reviewing the record's strengths and weaknesses. In particular, we need to understand how fossils provide a record of biodiversity.

The other important issue we need to come to grips with is time. Not the everyday time of daily use, the getting up at 6:30, work by 9:00, home at 5:30 and bed by 11:00 sort of time; I'm talking about billions of years of geological time (I prefer the evocative term 'deep time'). We will need to understand the way that geologists use deep time and the development of the Geological Time Scale. Dealing with the enormity of deep time does take some getting used to. There are some huge numbers involved. Life appeared on the planet about 3.7 billion years ago.[5] This is a staggeringly big number. Even a million is

---

[5] There are geochemical hints that life actually appeared before this, leading some workers to suggest that life reaches back beyond 4 billion years – an even bigger number!

hard to grasp – to help get the size of these numbers across to my first-year classes, I estimated how long it would take me to count to 1 million. The surprising answer was that it would take about 33 days of non-stop counting (no eating, no sleeping) to reach 1 million. To count to 3.5 billion would take around 290 *years* of non-stop counting.[6] I'll cover both of these topics in the next chapter.

[6] If you're interested in how I arrived at this estimate, I've outlined it in Appendix 2.

2 A Short Detour: The Fossil Record and the Geological Time Scale

The fossil record is the primary source of information that we can use to document the history of life on Earth. But it is a biased record, and although it's biased in a very predictable way, it does place limitations on the history we're trying to extract. We need to understand why the biases are there and see what can be done to minimise the problem.

Fossils are the remains of animals, plants, and other organisms preserved in rocks. They can be the body parts of animals, cellular material, tracks and trails left by ancient organisms, even faecal matter (although palaeontologists give that a better sounding name – coprolites). This record of the great panoply of life on Earth provides an astonishing amount of information. It can tell us about an organism's evolution, what extinct creatures looked like, how they behaved, and what sort of ecological setting they lived in. It documents how dinosaurs changed into birds, and how organisms evolved eyes – all this and more, much more. In this book, we are interested in the historical patterns of biodiversity, and for this sort of analysis, the fossil record is all we have available.

I like to compare the fossil record to a chain-link fence. The fence is strong, durable, and full of holes. The fossil record is strong: we have recovered many, many fossils. It's durable: no one (despite repeated efforts) has ever been able to prove it inconsistent. The same sequence of fossils appears everywhere on the planet – there are no rabbits in the Precambrian.[1] But like the chain-link fence, the fossil

---

[1] The story goes that someone once asked the British-born biologist J. B. S. Haldane what would disprove evolution. He reportedly replied: 'Rabbits! Rabbits in the Precambrian!' The Precambrian is a period of time between 540 million and 4.5 billion years ago – well before the evolution of bunnies.

record is full of holes, and it's important to understand why these holes exist.

There are three basic reasons:

(1) *Not all organisms (or parts of organisms) are capable of being preserved.*

Preservation in the fossil record is a chancy business. If I wanted to be included in the record – and really who wouldn't – then I would need to accept from the outset that my soft parts (skin, organs, etc.) are much less likely to be preserved than my hard parts – my skeleton. Indeed, fossils of all our hominin ancestors consist almost entirely of skeletal material. Occasionally, human soft parts are preserved – Ötzi the ice man died high in the European Alps about 5,500 years ago. He was frozen in alpine ice, which ensured that his body is complete; both his soft organs and bones are present; even his last meal has been documented. However, complete preservation such as this remains extremely unlikely, especially the further back in time you go.

The difficulty of preserving an organism's soft parts means that those with bodies that don't contain any hard parts *are unlikely to be preserved at all.* Bacteria and other single-celled organisms are affected particularly badly, but we also have a relatively poor record of leaves, flowers, and soft-bodied organisms such as worms, slugs, and jellyfish. This means that a significant proportion of life's biodiversity is either poorly represented or possibly not represented at all in the fossil record. Again, I need to emphasise that, against the odds, we do have many fossil remains of soft-bodied organisms. Some of the oldest fossils ever recovered, the 3.4-billion-year-old single-celled cyanobacteria from Western Australia and the wonderful 600-million-year-old Ediacara fossils, representative of some of the first complex multicellular life on Earth, are almost all soft-bodied. Nevertheless, it's clear that overall, the record is heavily biased towards organisms that have hard parts. This can be a result of either soft parts decaying away (or being eaten) or the fossilisation process itself, which often involves harsh chemical and physical changes that can damage or remove any or all of an organism's soft parts.

(2) *Not all environments are capable of preserving fossils.*

Continuing my quest to include what bits of me I can in the fossil record, I have to choose my final resting place extremely carefully. I need to pick a place where sediments are accumulating, not eroding. Dying on the top of a mountain would almost inevitably result in my remains being scattered (possibly by predators) and weathered away. A lake, swamp, or the sea would be the best. There I could be quickly covered by the accumulating sediment, keeping predators away. This rapid burial also has the advantage of shutting off the oxygen supply to my body, slowing down, or even stopping, the decay of my soft parts. Somewhere quiet would also be an advantage: no waves to roll me around and break up my mortal remains. So, lakes, swamps, and deeper marine settings are all good.

There are far fewer lakes and swamps than there are marine basins to accumulate sediment in, so the fossil record is massively biased towards marine fossils. This bias, coupled with poor preservation of soft parts, means that the vast majority of fossils recovered are of marine molluscs, clams and snails. As I said, it's a bias, but it's a predictable bias.

(3) *Species with low numbers of individuals and a limited geographical distribution are less likely to be preserved in the fossil record.*

Here I'm in real luck. I said earlier that the possibility of being included in the fossil record is dictated by chance. Let's say that in a certain species, owing to the difficulties of the fossilisation process and the limited number of environments that can easily preserve organisms, only 0.01% of the population is likely to be included in the record (this is probably an overestimate, but bear with me). If the species consists of 1 million individuals, then only 100 of them are likely to be candidates for preservation. In a population of 1 billion, 100,000 are candidates. Here am I, a member of *Homo sapiens*, a species that spans the globe with 7.8 billion individuals – this means I or one of my fellow humans have 780,000 chances to be preserved. I'll still need a lot of luck to be included in the record, but if I happen to be near a nice swamp, or out at sea at the appropriate time, I'm in.

The earliest body fossil, as opposed to geochemical evidence for life on Earth, is from rocks in Western Australia that are dated to about 3.7 billion years ago. These tiny fossils are of single-celled organisms, and some have been interpreted as prokaryotic cyanobacteria. Prokaryotes are the simplest of cells, lacking a nucleus to contain their DNA. They can, however, utilise a wide variety of energy sources in order to live, which in turn allows them to inhabit an extraordinary variety of environments. Some of the fossils from Western Australia are thought to be cyanobacteria, which means that, like plants, they make their living via photosynthesis. Prokaryotes are always single-celled – they never form colonies, let alone a multicellular body. These simple, but effective, organisms dominated the Earth's oceans for about the first 1.3–1.5 billion years of life's history.

In rocks about 1.5 billion years old, a group of organisms called acritarchs appear. Acritarchs represent the earliest forms of another major division of life, eukaryotes. Beyond these basics, we don't know exactly what these things are: the name literally means 'confused origin'. Eukaryotes are a major evolutionary advancement over the prokaryotes. They have a membrane around their DNA forming a nucleus. They also contain organelles. These are specialist cellular subunits, which are involved in several different activities within a cell, including protein manufacture and modification, energy production, and environmental sensing. The several types of organelles found in a eukaryote cell are enclosed in separate membranes. These organelles are thought to have evolved through a symbiotic relationship between prokaryote ancestors. Some organelles have their own DNA, inherited from their precursor prokaryote forms. Despite the added complexity, these early acritarchs are still single cells – but they don't stay that way. At about 600 million years ago, we have unequivocal evidence through the Ediacaran fossils that they have started to build bodies. Complex multicellular life has arrived.

Between the appearance of the first prokaryote cells at about 3.5 billion years ago and the eukaryotes building bodies at about 600 million years ago lies the majority of life's history – 2.8 billion years of it – and all the organisms of that period are soft-bodied. Because of the difficulty of preserving these forms, and the age of the rocks involved, they do not leave a good record. We do have fossils – some very good ones – but not enough to talk meaningfully about changes in biodiversity in the way I want to talk here. So, this book will mainly focus on the history of life beginning shortly before the arrival of the first multicellular life, about 700 million years of Earth's history.

The fossil record's strong bias towards preserving marine species raises a second issue. When life was confined to the oceans, this bias was clearly not important in interpreting changes in biological diversity. However, once life starts to encroach onto land, the fossil record becomes somewhat weaker, and any discussion of changes in terrestrial diversity can be a little problematic. Unlike the problems I outlined above in my discussion of the first 2.7 billion years of life's history, where I can say almost nothing, the restricted terrestrial record isn't a fatal flaw. There are significant numbers of fossils, and our understanding of them is growing.

BIODIVERSITY

The estimate of 8.7 million species on the Earth is a snapshot of the planet's current level of biodiversity. In this book, however, I'm interested in the changes in biological diversity over the millions of years of deep time. Instead of a snapshot of biodiversity, I'm interested in a video, showing how it fluctuates over the course of deep time. And that requires a different approach, one that takes time itself into account: we need to talk about rates. The level of biological diversity at any point in time is a result of a dynamic equilibrium between the rate at which new species evolve (also called the origination rate) and the rate that species become extinct. If the rate of extinction goes up and exceeds the rate of origination, then the overall level of biodiversity will fall. If the rate of origination exceeds that of extinction, the level of diversity will rise.

Today, most people think of diversity at the level of species: so many species per square kilometre, or so many species in that area of mangroves. However, because I am dealing with the uncertainties of the fossil record, where counting species can be difficult, I will often need to discuss biodiversity at a higher taxonomic level than species. This is particularly true when we get around to discussing mass extinctions in detail. A taxon is a population of organisms that are

Table 2.1 *Taxonomy of Humans*

Taxonomic hierarchy of Humans		
Kingdom	Animalia	
Phylum	Chordata	Animals with backbones
Class	Mammalia	Animals that have hair and suckle their young
Order	Primates	Humans, Apes, and Monkeys
Family	Hominidae	Humans and Apes
Subfamily	Hominini	Humans, Gorillas, and Chimpanzees
Genus	*Homo*	Humans
Species	*Homo sapiens*	Thinking man

grouped together because of their morphological similarities. A species is a taxon composed of similar individual organisms. Under the Linnaean system (named after Carl Linnaeus, who established the binomial method of naming species in 1735), similar species are grouped into genera, similar genera are grouped into families, and so on up the taxonomic tree to the highest taxonomic levels, phylum and kingdom.[2]

Table 2.1 shows a basic taxonomic hierarchy for humans, *Homo sapiens*. We need to note a couple of things about this table. Firstly, it doesn't include fossil species. If it did, *Homo sapiens* wouldn't be the only species included in the genus *Homo*. We would need to add up to eleven others, including *Homo neanderthalensis*, *Homo erectus*, *Homo antecessor*, *Homo naledi*, and *Homo floresiensis*.

Secondly, this ladder of taxa isn't just thrown together at random; it is arranged to reflect evolutionary relatedness. All the mammals share a single common ancestor in the animals with backbones. All the primates share a single common ancestor in the

---

[2] There are other taxonomic systems, some perhaps more useful than the Linnaean, but most of the literature I'll deal with here is based on the Linnaean system, so I will continue to use that.

mammals. Apes, humans, and monkeys (the family Hominidae) share a single common ancestor within the primates and so on. Out of interest, and to show how closely related to gorillas and chimps we are, I've shown the subfamily level (Hominini) – we get down that low in the hierarchy before humans are considered to be separate from the great apes. Some biologists have suggested that we are so similar to the gorillas and the chimps that they should be included with us in the genus *Homo.*

ORIGINATION AND EXTINCTION

Origination is the word some palaeontologists use for what biologists call speciation – the evolutionary appearance of a new species. Palaeontologists also have a different method of defining exactly what a species is. The biological definition of a species is *a group of interbreeding organisms that are reproductively isolated from other groups.* That is not a very useful definition for a palaeontologist; put two fossils together and they are never going to reproduce, whether or not they are the same species. Palaeontologists use what is known as the morphological definition. Essentially, that means if two fossils look very similar, they are considered to be the same species. This may seem a little rudimentary, but using the biological definition of a species also brings together individuals that basically look the same (allowing for geographical variants and the differences between females and males in some species).

Understanding how new species evolve is at the heart of Charles Darwin's great book *On the Origin of Species by Means of Natural Selection.* A common claim today is that although Darwin provided arguably the most important mechanism for evolutionary change (natural selection) and had explained wonderfully the process of adaptation, he failed to demonstrate how new species evolve. I think this claim, if not wrong, is not telling the full story. Darwin didn't have the access to the scientific and technical advances that we have available, notably our understanding of genetics that tells us more about the process of speciation. Nevertheless, he showed that through the agency of natural selection, an individual organism will accumulate small morphological changes, which, as they spread through a population, will result in a gradual divergence from the ancestral stock. Over time, the original and the diverged forms will become reproductively isolated – unable to interbreed – and a new species appears in the fossil record.

The extinction of a species occurs when the last individual belonging to that species dies. Some species seem to disappear slowly, away from the glare of publicity. We don't notice until they are gone. In some cases (unfortunately, an increasing number of cases), we know precisely when a species goes extinct. The extinction of the passenger pigeon (*Ectopistes migratorius*) occurred on 1 September 1914 when the last specimen, a female called Martha, died in the Cincinnati zoo. It took a lot of effort to kill off this species. In the nineteenth century, one in every four birds in North America was a passenger pigeon. They were gregarious birds, travelling in huge flocks. In 1866, one flock was estimated as being 1.5 kilometres wide and 500 kilometres long. Contemporary reports claim that it took 14 hours to pass.

The passenger pigeon's extinction was a direct result of the expansion of European colonists across the United States. The pigeons were a cheap food source for the hungry poor and were massacred on an industrial scale. The last confirmed sighting of a passenger pigeon in the wild was in 1902. When Martha died 12 years later, the species was no more. To commemorate the 100 years of passenger pigeon extinction, in 2014 the Smithsonian Museum put Martha's stuffed remains on public display for the first time (prior to this commemoration, she had been considered too valuable to risk displaying). It's small comfort that her remains formed the centrepiece of a display warning of the dangers of extinction.

Today, stories like that of the passenger pigeon are all too common as human activity destroys natural habitats and changes the climate. The finality of extinction is brought home to us almost daily in the media, so it is perhaps surprising to realise that up until the late 1700s the fact of extinction wasn't recognised at all. The natural scientists of the time considered the planet and all its creatures to be part of God's perfect creation. Having created it, why would he allow parts of it to die away? Even at that time, it was clear that humans had driven some animals to extinction. Perhaps the most famous was the dodo. The last confirmed sighting of this much-maligned bird was in 1662. But this wasn't the first extinction that humans were directly responsible for: the moa in my adoptive Aotearoa/New Zealand were hunted to extinction well before the last dodo, and we will see there are even earlier extinctions in which humans are implicated. But all these extinctions were brought about by the hand of man, not God – they didn't challenge the acceptance of the perfection of creation.

The emerging science of geology,[3] and in particular palaeontology, the study of fossils, did provide that challenge. Many of the fossils that scientists were studying were clearly the remains of creatures that did not seem to be the same as creatures that currently roamed the Earth. At the time, the argument went, that didn't matter: the world was a big place and much remained to be explored, so the animals that produced these fossils would be located – eventually.

It was Jean Léopold Nicolas Frédéric Cuvier (known by all, for reasons that escape me, as Georges Cuvier), the great French comparative anatomist, who established beyond doubt the fact of extinction. Cuvier, a survivor if ever there was one, prospered through the various stages of the French Revolution and during Napoleon's reign worked in the great *Muséum national d'Histoire naturelle* in Paris. The museum was collecting huge amounts of fossil material from around the world, and the nearby zoo offered a stream of just-dead bodies to compare with this fossil material. No one else had access to such material, and Cuvier used everything he was presented with.

---

[3] It would probably have been called physical geography at the time. For a detailed account of the emergence of historical geology as a separate discipline, you can't do better than Martin Rudwick's monumental two-volume history, *Bursting the Limits of Time* and *Worlds Before Adam*.

In his fossil material, Cuvier saw animals that – although similar – were not identical to today's. He worked on fossils of mastodons that, while clearly related to today's elephants, were not elephants. The unexplored areas of the globe were getting smaller, so it was becoming increasingly unlikely that an elephant-sized animal like a mastodon had been missed (although the hope that a living one would be found in the great prairies of North America lingered on until very late). Cuvier concluded that the fossil mastodons represented a species of animal that was no more: they had gone extinct.

Cuvier was also a champion of the new continental idea of looking at fossils and using them in the same way an archaeologist uses Roman artefacts to build up a history of a Roman town. He used fossils to establish the ancient (he called it pre-Adameic) history of an area. Cuvier was one of the first scientists to write down what we would now call a geological history. To me, extinction appears central to this notion. The fact that there had been animals on Earth that were not the same as animals today raises an obvious question: what happened between the time that animals we find as fossils walked the Earth and today's faunas? And that question demands a historical explanation.

EXTINCTION, ORIGINATION, AND THE GEOLOGICAL TIME SCALE

Regardless of its initial lack of recognition, extinction was unwittingly being used by geologists of the past as a tool to order the outcrops of rocks they were studying. In the 1830s, Adam Sedgwick studied the rocks of Devon and, because they contained an assemblage of fossils that weren't found in rocks that lay above or below them, he grouped them into a geological unit called a system that he named after Devon – the Devonian. He grouped some of the rocks of Wales into another system, which he named the Cambrian, after the Roman name for Wales, Cambria. At about the same time, the splendidly named Sir Roderick Impy Murchison examined other Welsh rocks and erected the Silurian system, named after an ancient Celtic tribe from the area, the Silures. The system of rocks he recognised around Perm in Russia he named the Permian. Geologists were identifying their rock systems based on the unique fossil faunas each contained. These faunas were unique because they were based on an underlying pattern of extinctions and originations.

Early geologists soon realised that the fossil faunas that defined each rock system could be recognised over a wide geographical area. This in turn allowed the rock systems to be recognised over the same wide area. It also became clear that the rock systems and the fauna they contained succeeded one another in time. This allowed the various systems to be listed in chronological order. Careful field studies of the relationships between these geological systems (especially the way they were stacked vertically in outcrops) showed that rocks of the Cambrian system were older than the Silurian system rocks (with another system, the Ordovician, in between); the Devonian system is younger than, and sits right on top of, the Silurian. The rocks of the Permian system are younger than any of these.

The recognition and naming of geological systems, and importantly their ordering in time, ultimately led to the development of the present-day Geological Time Scale, part of which is shown in Figure 2.1 (the full scale is provided in Appendix 1). Note that what I referred to as a *system* in the preceding discussion on the time scale is double-labelled in Figure 2.1 as a *period*. This is because geologists maintain two different, but parallel, sets of terms, one when we refer to the rocks themselves (system), the other when we talk about the time when the rocks were deposited (period). So, the Ordovician system of rocks was deposited during the Ordovician Period of time. This separation of time and rocks is meant to add precision to dialogues between geologists. However, to some geologists (myself included), the maintenance of two parallel systems appears to be a pedantic complication. Currently, there is a significant amount of discussion in the geological community about just using one set of terms – the time units. I'll follow that notion here and talk about the rocks deposited in a particular period.

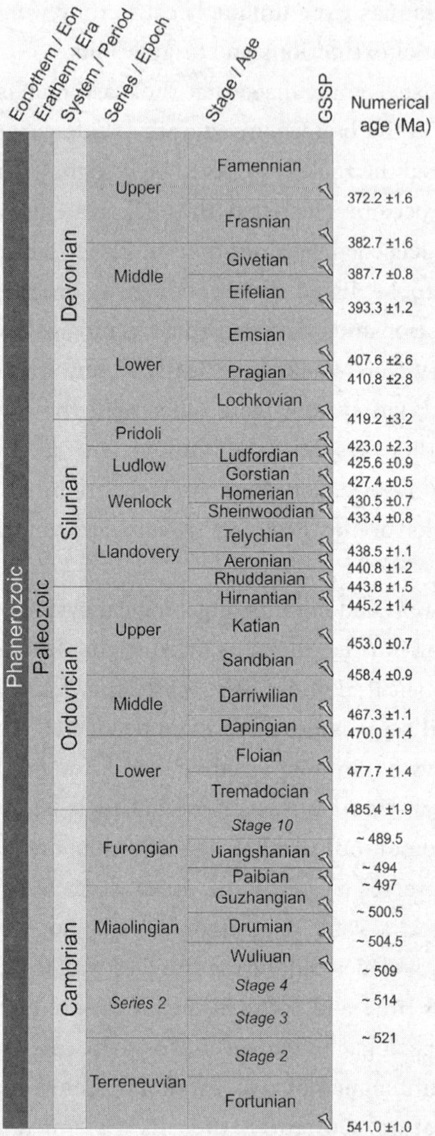

FIGURE 2.1 Part of the modern Geological Time Scale. For the complete time scale, see Appendix 1 (Cohen, Finney, Gibbard & Fan, 2013; updated). Reproduced by permission of the International Commission on Stratigraphy.

Today's Geological Time Scale is a far more sophisticated affair than the original efforts of Sedgwick, Murchison, and their colleagues. Like the taxonomic tree we saw earlier, it's hierarchical. Periods are subdivided into epochs, then ages. All the periods shown in Figure 2.1 are grouped into the Paleozoic Era. In this book, I will concentrate on three categories: era, period, and epoch. Regardless of the increase in complexity of the time scale, it is constructed using the same basic tool as those early geologists. Underlying the modern time scale is a pattern of extinctions and originations that make up a well understood sequence of fossil assemblages. After much discussion on what was the correct faunal sequence, geologists agreed on the ordering of the various geological time units. Once the order was agreed on, it takes a lot to change it. Recently, some changes have been made to the definitions of some periods (and new ones created) that will affect some of my discussions, especially in Chapter 3 where I'm going to talk about the appearance of early animals and the emergence of the Earth System.

On the far right-hand side of Figure 2.1 are the current measurements of the ages of the boundaries between the periods, stages, etc. These are not given as millions of years but as Ma – mega annum; 1 mega annum = 1 million years before the present day.[4] Unlike the agreed order of the geological periods, which is hard to change, the numerical ages of the boundaries between geological time units can be and indeed frequently are changed. This is because techniques used to date rocks are constantly improving, producing ever more accurate estimates of ages. As the new ages are released, they are incorporated into the time scale. Happily, with every update of the time scale, the shifts in the ages have become less and less. We can be very confident of these age measurements.

I have spent some time talking about the Geological Time Scale. This is because the time scale is the basic tool of geology,

---

[4] To confuse things a little further, we use millions of years (Myr) when we talk about duration. For example, the Cretaceous Period lasted for about 79 Myr.

forming the basis of the way geologists and palaeontologists communicate with each other. I'll refer to it a lot as we move on.

RECOGNISING ORIGINATION AND EXTINCTION IN THE FOSSIL RECORD

The recognition of origination and extinction in the fossil record is not entirely straightforward. Palaeontologists collect rock samples and make a record of the species they contain. An origination event – the evolution of a new species – is placed at the oldest sample in which the new species appears, often called the *first appearance datum* or FAD. The extinction of a species is placed in the youngest sample containing the species. In a fit of consistency, that's called the *last appearance datum* or LAD. Figure 2.2 sets out the situation for three fossil species. Down the left-hand side is the rock section being examined, together with its age in millions of years. Note that the diagram is plotted against the thickness of the section, so the ages in millions of years are not regularly spaced; this reflects the differing rate at which the sediments accumulated. The next column shows the eleven samples the palaeontologists collected and analysed. The next three columns make up what is called a *range chart*. It's a listing of which of the three species were found in each sample. Their presence in a sample is indicated by a different symbol.

Look at species 1. It has its FAD in sample 1 and its LAD in sample 7. It has a temporary absence in sample 4, but since it reappears in younger samples, we can ignore this. I've used the same logic to establish the FADs and LADs for the other two species. Does this mean that we are documenting, with certainty, a species' evolution, and extinction? Unfortunately, no. We have already seen earlier in this chapter that a species can be absent from the fossil record for a number of reasons. In the case of species 1, it's possible that environmental conditions changed between samples 7 and 8, and the species wasn't able to be preserved in the younger samples, or the species couldn't live in the changed environment although it was happily existing elsewhere. So, although the data suggest that species 1 went

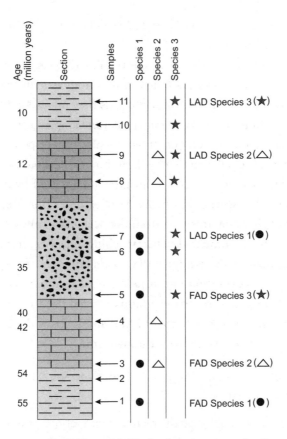

FIGURE 2.2 FADs and LADs for three imaginary fossil species. The
vertical axis is an imaginary section of rocks with its age in millions of
years recorded to the left. Plotted alongside are the sample locations,
marked by arrows and the record of fossils recovered from
each sample.

extinct in sample 7, this could simply be where it stopped being
preserved or mean that it couldn't tolerate the new environment,
resulting in its absence in the younger samples.

Sampling issues can also confuse the issue: the FAD of species
1 is in the oldest sample and the LAD of species 3 (the grey stars) is in
the youngest sample. How do we know that if we could have collected
samples from further down in the section or even higher in the section,
we would not have found more specimens of species 1 and 3? The short

answer is that we can't be sure. These FADS and LADs could just be an artefact of limited sampling.

Is there a solution to this uncertainty? Yes, there is – more data. Palaeontologists collect and examine samples from as many sections as they can lay their hands on. These sections may represent different depositional environments, some may be better sampled than others, and some may contain well preserved fossils, others poor. These sections, when stitched together, will improve the quality of the record. A poorly sampled section from here is enhanced by a better one from there; this section has many more samples from deeper in the record than that one; this core from the deep ocean allows a very long record to be documented. This stitching together of sections across both time and space allows the documentation of a species' full range – from its evolution to extinction, or as near to it as we can get. There will always be some uncertainty around the range of some species, but the resulting catalogue of FADs and LADs is the basis for any discussion of diversity changes through time.

THE SEPKOSKI DATA SET AND THE BEGINNING OF MODERN EXTINCTION STUDIES

In the 1700s, Cuvier established the reality of extinction. As our understanding of extinction increased, and more details of the fossil record were uncovered, geologists noted intervals of increased levels of extinction, for example at the Cretaceous–Paleogene boundary, the point in time where the dinosaurs went extinct. Palaeontologists started to investigate what caused these sudden apparent increases in levels of extinction. Driven, I suspect, by public fascination, much of the discussion about what happened at the end of the Cretaceous focused on 'what killed the dinosaurs'. There was a range of suggested dinosaur-ending solutions – and we'll talk about some of them later in the book.

I think that modern, systematic, and quantitative studies of extinction and mass extinction started with the compilation of what is commonly called the 'Sepkoski data set' in the 1970s. My decision to use this starting point is arguable. Palaeontologists such as Norman

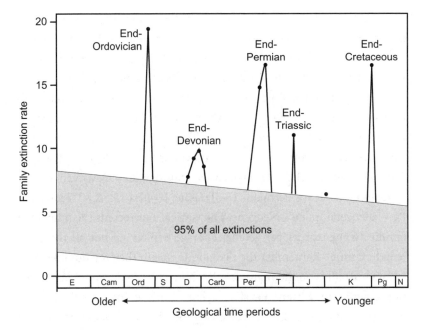

FIGURE 2.3 The Big Five mass extinctions in stark relief. The vertical axis is the number of families going extinct in each of the geological stages, which represent varying lengths of deep time. The horizontal axis is time. The symbols represent the various geological periods, Cambrian, Ordovician and so on; see text for details. Adapted with permission from Raup and Sepkoski (1982).

Newell were discussing mass extinctions in 1962, and before that, in 1950, Otto Shinderwolf had speculated that the dinosaur extinctions were the result of a burst of cosmic rays caused by a supernova. But the development of the Sepkoski data resulted in a leap in our understanding of patterns of extinction; to me, it's an obvious place to start. The data set, compiled by Jack Sepkoski at the University of Chicago, is a global compendium of the FADs and LADs of many marine genera and families per geological period (note that we've slipped up the taxonomic hierarchy from species). Compared with modern data sets that I'll talk about later, there is nothing particularly sophisticated about the Sepkoski data set, but it started a revolution in the way we think about extinction. In 1982, Jack Sepkoski joined with David Raup and published a seminal diagram that I have redrawn in Figure 2.3.

Along its horizontal axis is 600 million years of geological time (see Appendix 1). This is not the same time scale as the one used by Raup and Sepkoski; I have updated the millions of years to reflect the latest estimates. In addition to the millions of years, I have also shown the relevant geological period. This is the same time scale that I will be using elsewhere, so it's worth spending some time on it here. Again, a detailed version of the 2020 time scale is in Appendix 1. The letter codes represent the various geological periods: E = Ediacaran, Cam = Cambrian, Ord = Ordovician, S = Silurian, D = Devonian, Carb = Carboniferous, Per = Permian, T = Triassic, J = Jurassic, K = Cretaceous, Pg = Paleogene and N = Neogene. The vertical axis records the number of families going extinct per geological stage (which are not all the same length of time). Remember the taxonomic tree in the previous chapter – families usually contain many species. Ninety-five per cent of all extinction events fall within the shaded box – this is, of course, the vast majority of the data. However, sticking their heads up above the shaded area are five outliers, intervals of exceptionally high rates of extinction – events that have come to be known as the Big Five mass extinctions. Mass extinction had been widely discussed, and the term was in relatively common currency when Raup and Sepkoski published their work, but here they are, stark and unambiguous, for all to see. It was now possible to talk about their timing and magnitude with some degree of certainty.

The diagram essentially splits extinction into two types:

(1) *Background extinctions.*

The standard run-of-the-mill extinctions caused by the intense competition for survival in the wild – Nature really is red in tooth and claw. Darwin saw this form of extinction as the driver for natural selection. Figure 2.3 suggests that background extinction represents perhaps over 95% of all extinctions. Figure 2.4 shows what background extinction is supposed to look like in the fossil record – a steady turnover of species over time, driven by the Darwinian notion of competition between individuals for limited resources.

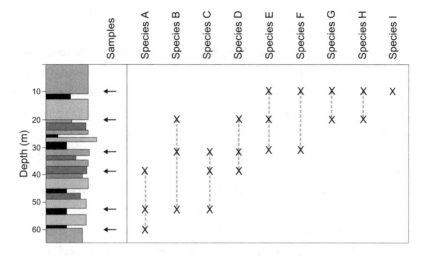

FIGURE 2.4 A cartoon of a range chart showing a typical pattern of background extinction.

(2) *Mass extinctions.*

Short, sharp, shocks representing periods of heightened levels of extinction superimposed on background extinction. It has been said that a soldier's life is one of long periods of boredom separated by short intervals of terror. Figure 2.3 suggests that the same thing is true for life on Earth.

Today, the mass extinctions recognised in the Sepkoski data set are commonly referred to as the 'Big Five'. They are, in descending order of age:

- End-Ordovician (about 444 million years ago);
- End-Devonian (about 370 million years ago);
- End-Permian (about 250 million years ago);
- End-Triassic (about 200 million years ago);
- End-Cretaceous (about 66 million years ago).

These mass extinctions have become immortalised in both the scientific and popular press as the Big Five, with the current biodiversity crash often referred to as the 'Sixth Extinction'. For reasons outlined in Chapter 1, when talking about our current biodiversity crisis I prefer the term defaunation.

We should also note that the mass extinctions identified by Raup and Sepkoski all fall at, or very near to, the boundaries of geological periods. Given how the Geological Time Scale was developed, this should be no surprise. Geologists placed the period boundaries at levels of significant change in fossil assemblages – and what could cause more change in the assemblages than a mass extinction?

As an aside, look at the way the shaded area in Figure 2.3 (containing 95% of the extinction data) slopes downwards from left to right. This suggests that the rate at which families went extinct in, say, the Ordovician is higher than in the younger periods, such as the Neogene. A more recent analysis of extinction data suggests that this slope may be real. This implies that over the past 600 or so million years, the likelihood of a family going extinct has reduced – it's getting easier to survive on Earth. Not everyone accepts that this is the case, and those that do don't really have an explanation of why it might be so.

Having covered the basics of the fossil record and its limitations, we can move onto the history it contains. We are starting that history with the origin of multicellular animals and the emergence of the 'modern' Earth System about 700 million years ago. This is a story that is intimately connected to the complex history of oxygen on Earth.

3   The Origin of Animals and the Emergence of the Earth System

INTRODUCTION

A fully operational Earth System didn't happen overnight. It needed the evolution of multicellular organisms and the appearance of fully functioning modern-style ecosystems. Multicellular organisms first became reasonably common in the fossil record some 600 million years ago. We've met them already: the Ediacara fauna. Proper ecosystems with complex relationships between predators and prey took a little longer to appear. They are first found in the record at around 540 million years, near the beginning of the Cambrian, and are associated with the beginning of a massive increase in the planet's diversity.

This period of Earth history – as single-celled organisms develop into multicellular organisms, fundamentally changing the planet and heralding the appearance of the modern Earth System – is fascinating. It's a period of complex physical, geochemical, and biological changes, and in this chapter, I can only provide the briefest outline of the story. But before I do that – I think a field trip is in order.

BRACHINA CREEK

The Flinders Ranges in South Australia are arguably that state's only significant mountain chain. Even then, the peaks are not very grand, the loftiest being a mere 1,170 metres (about 3,840 feet) high. The ranges strike out from the top of the Spencer Gulf and stretch out northeast for about 400 kilometres (about 250 miles). They are narrow, discontinuous, and rise surprisingly fast from the surrounding flat, salt-lake-studded plains that seem to stretch to infinity.

The rocks that make up the ranges are part of the Adelaide Superbasin (formally known as the Adelaide Geosyncline), a basin within which sediments were deposited during the Neoproterozoic Era and Cambrian Period. These sedimentary rocks contain an astonishing record of the appearance of complex life. In some areas, the rocks have been twisted and deformed as older material has punched its way towards the surface, forming structures called diapirs; in others, the layers of rock dip gently towards the west, and gorges through the ranges offer easy access to the history recorded in them.

Brachina Creek cuts east–west through one of these areas of simple deformation. The creek has cut spectacular gorges, which attract many tourists each year. They can follow a geological trail; signs and leaflets are available to help visitors understand the geology they are passing through. On my latest visit, I didn't need these signs or leaflets, as I was on a trip led by Jim Gehling from the South Australian Museum, one of the world's leading experts on Ediacaran fossils and the geology of the Flinders Ranges.

Traffic along the creek is one-way from east to west, and that was the way Jim led us. Because the layers of rock dip into the Earth's surface, this direction of travel meant that Jim's trip started in older rocks, and as we travelled along the creek the rocks got progressively younger. That's the way a geologist would tend to describe a sequence of layered sedimentary rocks, but I think it's easier to tell the story if I start with the younger material in the west and move backwards through older rocks as we progress eastwards. I have shown the route we took (diagrammatically) in Figure 3.1. The thick black line represents the ground surface, with the various rock units shown dipping below it. The younger rocks are to the left of the diagram, older to the right. Along the route the important features we will encounter are shown. Across the top I have indicated the age of the rocks at the surface in terms of the geological periods we will pass through.

Looking away to the west from the beginning of our trail, there is no real topography, just a flattish emptiness with scrubby vegetation, with no rock outcrops to be seen. The ranges proper rise behind us. Around us the topography is subdued, covered with small

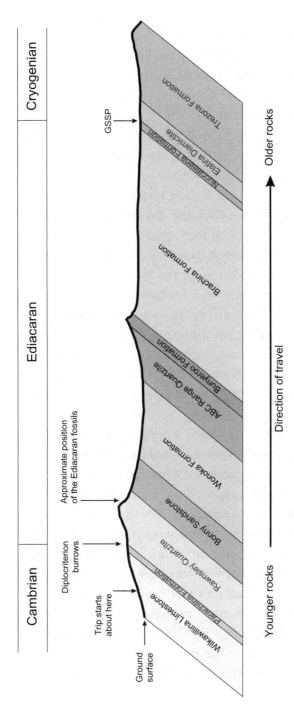

FIGURE 3.1 The field trip route along Brachina Creek. The distance covered on the trip was about 8.5 kilometres (a little over 5 miles).

eucalyptus trees. Along the dry creek bed are outcrops of early Cambrian rock unit called the Wilkawillina Limestone that contains fossils of a sponge-like animal called *Archaeocyatha*. Fossils of these organisms were some of the first fossils I ever collected as a child – so I have a soft spot for them.

Moving eastwards along the creek, we start our descent into deep time. The rocks around us change to fine-grained silts and sands formed on the edge of an ancient sea. Because these siltstones and sandstones are softer than the limestones around them, they are more easily eroded, and the countryside flattens out. Geologists have named these rocks the Parachilna Formation. The base of the Cambrian Period (see the time scale in Appendix 1) in the Flinders Ranges has been placed at the very bottom of this formation, and it is marked by a layer containing an abundance of what are known as trace fossils: the remains of burrows, trails, and tracks. Trace fossils can record (among other things) organisms moving, feeding, mating, and fighting. In other words, they allow us to document the everyday life of an extinct animal. Frustratingly, it is very rare to be able to link a trace fossil from this formation with the organism that produced it. In the Parachilna Formation, the base of the Cambrian is marked by an abundance of U-shaped burrows called *Diplocriterion*. Passing by these fossils, as we go back further in time, we enter rocks that are older than the Cambrian, a relatively recent addition to the geological timescale, the Ediacaran Period.

The topography ahead rises steeply as we approach the hard rock units of the Rawnsley Quartzite. From the western side is a steadily rising slope, but it terminates in great, precipitous, eastward-facing bluffs. During the day, in bright sunlight, these bluffs look a bruised blue-purple, but with the rising and setting Sun they glow a deep ochre-orange. Brachina Creek has cut deep gorges though the Rawnsley Quartzite, exposing sandstones that accumulated in shallow seas over 540 million years ago. Jim's tour stopped in the middle of one of the gorges, and an excited group of palaeontologists tumbled out of the van to search for the Ediacara fauna that was first identified in the Flinders Ranges. Jim had assured us that this particular location was an excellent place to find the fossils, so we struggled up a steep talus slope to find our prize – but to no avail; no Ediacaran fossils were found that day. Disappointed, we laboured back into the van and moved onto the next outcrop.

Continuing our own trip backwards in time, we leave the gorges behind and move into more open country as we pass through more soft siltstones and sandstones. There are no body fossils in these rocks, just a few trace fossils and some algal remains called stromatolites . As we travel further back in time, even they disappear. Pressing on, we pass through a set of ridges that mark the presence of a more durable layer of sandstone (the ABC Range Quartzite). Beyond this hard unit we reach a unit of red-brown siltstone, the Brachina Formation. It's a thick unit, and at the very base of this unit we come across something quite different: the Nuccaleena Formation. This formation is only a few metres thick: it's a dolostone, composed almost entirely of a carbonate mineral called dolomite. The sandstones and siltstone we've been travelling through were sediments that were washed down rivers and deposited on the sea floor. Dolomite is different: it's a chemical pre-cipitate that under the right conditions forms directly from sea water. Immediately underlying the Nuccaleena Formation is another unusual rock unit, the Elatina Diamictite.

A diamictite is a rock with large angular pieces of older rock surrounded by a very fine matrix. Diamictites like this are deposited by glaciers. Glacial deposits of this age (about 635 Ma) are not uncom-mon and have been identified in locations widely scattered around the world, indicating that the world had experienced a major ice age. During the peak of the last ice age, about 20,000 years ago, ice extended out from both poles. In the Northern Hemisphere, an ice sheet covered Canada and extended down into the northern United States, and much of Europe was also under ice. The ice sheet that deposited the Elatina Diamictite over 630 million years ago was much bigger. Evidence suggests that the ice extended from the poles all the way into the tropics. Essentially the Earth was frozen like a gigantic snowball from the poles possibly right down to the Equator (hence the term 'Snowball Earth')[1]. This ice age is named the Marinoan Glaciation, and it lasted about 10 million years.

The Marinoan Glaciation wasn't the only Snowball Earth event. There's an older one also recorded in South Australia, the Sturtian[2] Glaciation, which ended about 680 million years ago. There is also some equivocal evidence, largely from Africa, of an even older Snowball Earth event. Missing from the younger sequence of rocks exposed along Brachina Creek, there is evidence elsewhere of a much younger (581 million years old) snowball event, which represents a brief return to a global ice age. Called the Gaskiers Glaciation, it lasted about a million years. The three older glaciations, the Sturtian, Marinoan, and African Snowball Earth events are all included in another new geological period, the Cryogenian. The Gaskiers Glaciation is in the younger Ediacaran Period.

For every boundary between periods, a *type section* is defined to provide a standard reference for studies, and a 'golden spike' (well, bronze, actually) is hammered in and firmly glued, to mark the exact placement of the boundary. This 'golden spike' also has the less glamorous name of Global Standard Section and Point, or GSSP (marked on the time scale in Appendix 1). Jim Gehling and his colleagues worked very hard to get the section in the Flinders Ranges declared as the international type section for the base of the Ediacaran, and in 2004 they succeeded. Our trip terminates at the

---

[1] One model for what triggered the snowball event suggests that because at the time all the continents were confined to warm and wet tropical latitudes, chemical weathering ran rampant. This depleted the level of carbon dioxide in the atmosphere, significantly reducing the greenhouse effect – and the planet started to freeze. The planet melted only when the ice cover stopped the chemical weathering and volcanoes were able to increase the level of atmospheric carbon dioxide, restoring milder climatic conditions.

[2] Both the Marinoan and Sturtian glaciations were first recognised by the Antarctic explorer/geologist Sir Douglas Mawson in rocks from the Adelaide Geosyncline. He recognised their glacial origins but not their global significance. The ages of these glaciations are somewhat uncertain, and several estimates exist in the scientific literature; I have taken these dates from the book *The Cambrian Explosion: The Construction of Animal Biodiversity* by Douglas Erwin and James Valentine. See the Further Reading section.

Ediacaran GSSP. Visiting one of the newest type sections was one of the highlights of the field trip for our group of geologists, and, with a degree of ceremony, a cloth and some cleaner were produced and the golden spike was carefully polished. In other countries, when a type section is awarded to them, a great fuss is made. The type section of the Permian–Triassic boundary is in China, and to celebrate, the government developed monuments, gardens, and an interpretive centre to help geo-tourists understand the importance of the place. The Australians are more phlegmatic and have a galvanised metal post[3] to point the way!

Now let me tell you why this field trip is both exciting and important.

THE EDIACARA FAUNA

Our field trip covered the late Cryogenian to early Cambrian periods, an interval of time spanning approximately 530 to 640 million years. After spending the better part of 3 billion years as single-celled organisms, in the 170 million years or so between the beginning of the Sturtian Glaciation and the early Cambrian, life takes off, and multicellular animals make their first appearance in the fossil record. The oceans fill with new species, slowly at first, then – near the base of the Cambrian – in a vast explosion of forms, coincident with the first appearance of a modern-style ecosystem. Figure 3.2 presents a broad outline of some of the key events that took place between the Sturtian Glaciation and the Cambrian Period. Down the vertical axis is the time scale, in millions of years (Ma) and corresponding geological periods. The three horizontal shaded bands are the three Snowball Earth glaciations mentioned above. Most events shown are biological, but down the right-hand side of the diagram I've put some indication of the level of oxygen in the ocean – I'll come back to that later in this chapter.

Our first hint of the presence of reasonably simple animals on the planet is from rocks in Oman that pre-date the Sturtian Glaciation.

---

[3] It's metal to avoid it being eaten by termites.

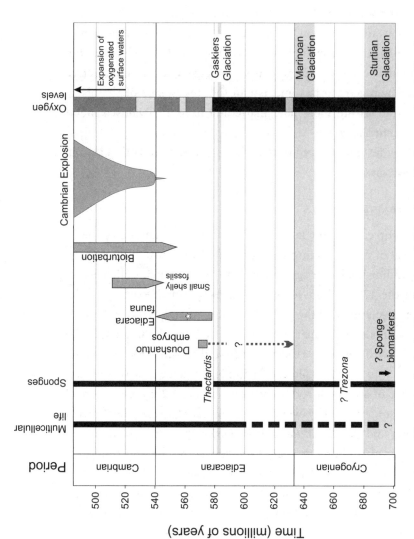

FIGURE 3.2 The origin of animals. Time down the left-hand side. The horizontal shaded bands are snowball Earth events. The column on the far right of the diagram represents oceanic oxygen conditions. Black, low oxygen; dark grey, increased levels of oxygen; light grey boxes, period of heightened levels of oxygen. For details see text. Animal data from several sources listed in the additional reading. Oxygen data from Cranfield (2014) and Wood and others (2019).

These are not what you would call traditional fossils: they are biomarkers or chemical fossils. As part of the process of living, many organisms produce complex organic molecules (usually lipids) that are unique to the organism that is producing them. Despite all the difficulties of fossil preservation discussed earlier, these molecules are often preserved, and the rocks from Oman contain the biomarkers that were probably produced by sponges. There is no fossil of the actual organism preserved, but the chemical trace of the sponge's presence remains. These biomarkers are shown at the base of the line marked sponges towards the left-hand end of Figure 3.2. It's no surprise that sponges, the simplest of all animals, are the first multicellular animals to appear on the planet.

If, on our field trip, we had gone on, deeper in time, past the Ediacaran GSSP, we would have reached another rock unit called the Trezona Formation (Figure 3.1) that lies between the Marinoan and Sturtian glacials. This unit contains abundant algal remains called stromatolites. In 2010, a possible fossil sponge was described from this formation. Named *Trezona* after the rock unit it was found in, it is controversial, and no one is all that confident about definitively calling it a sponge (some think the fossil is a result of ancient mud cracks), but many workers give it the benefit of the doubt.

The only way to get a three-dimensional impression of the fossil is to grind the rock that contains it down millimetre by millimetre and draw the fossil as it appears on each ground surface. The series of drawings is then assembled into a three-dimensional model.

The oldest fossil of an actual sponge that we are certain of is *Thectardis* from Mistaken Point in Newfoundland, Canada. It appears in the record just after the Gaskiers Glaciation. However, by this stage we have excellent and abundant evidence for multicellular animals, including the Ediacara fauna and some remarkable fossils recovered from the Doushantuo Formation in the Guizhou Province of South China.

Fossils from the Doushantuo Formation are all marine, microscopic, and above all, beautifully preserved. The fossils are small, about 500 micrometres in diameter, and are gently coaxed out of the rock in acid baths. Most workers agree that these are fossils of animal embryos. A few go as far as recognising various stages of embryonic development in them and separating them into embryos of sponges and bilaterians.[4] There is a lot of uncertainty surrounding the age of the embryos in the Doushantuo Formation. Most of the fossils seem to just postdate the Gaskiers Glaciation; however, the formation itself extends down almost to the Marinoan Glaciation, and there are reports of embryos from lower in the unit. I have indicated this uncertainty with the dashed line and question mark in Figure 3.2.

Also appearing very soon after the Gaskiers Glaciation is the Ediacara fauna itself. I've mentioned this assemblage of fossils several times before, and I make no secret of the fact that these are some of my favourite fossils. I did my geological training at the University of Adelaide in South Australia where I had access to many specimens, and I remember staring at the strange disks and fronds with delight. I was occasionally involved in trips with fellow students to the Flinders Ranges, looking for new specimens. We only looked, never collected: the fossils are protected by law, and unauthorised collection is illegal. Nevertheless, I can still remember the thrill of finding some of these astonishing fossils in the field.

The Ediacara fossils were first discovered in 1947 by the geologist Reg Sprigg in the Ediacara Hills of the Flinders Ranges of South Australia. They have now been found in rocks of a similar age from around the world. Palaeontologists who originally worked on the Ediacara fossils assigned these admittedly odd-looking fossils to existing animal groups. A segmented organism with a definite head end (called *Spriggina* after Reg Sprigg) was considered to be an ancient annelid worm. *Mawsonites* looked for all the world like a fossil jellyfish. Other fossils were just odd. For example, no one was sure exactly what to make of a dome-shaped circular fossil known as

---

[4] Bilaterians are animals with bilateral symmetry, implying the presence of a mouth, gut, and anus.

Tribrachidium, only 1–2 centimetres (slightly less than an inch) in diameter with three clear hooked ridges. We still don't know – one thought is that it represented a very ancient echinoderm.[5]

In 1992, the German palaeontologist Adolf Seilacher suggested that the Ediacaran animals represented a failed evolutionary experiment, and that none of the Ediacaran forms were in fact related to any of the animal groups that exist today. He called them Vendobionts[6] and believed that they were so different to today's animals that they needed their own kingdom. Although Seilacher's idea is still being discussed, most workers now agree that although the Ediacaran animals were a little weird looking, they don't need a separate kingdom, and many are recognised as ancestors to today's animal groups. However, most of the disk-shaped organisms like *Mawsonites* that were originally described as jellyfish appear now to be attachments used to anchor frond-shaped animals to the sea floor.

In 2020, a new but very important Ediacaran fossil was recorded from an area called Nilpena in the Flinders Ranges. Nilpena is an astonishing location. To protect the fossils found there from looters, the area has been fenced off and can only be entered though a secure gateway. Within the fenced area, palaeontologists have literally uncovered the ancient Ediacaran sea floor. Large sheets of rock, often many square metres in area, preserve a very precise copy of the sea floor as it existed over 550 million years ago, with all its microbial mats and fossils of the grazers that feasted on them. There, amongst the microbial mats, a team of palaeontologists led by Scott Evans from the University of California located a tiny (often less than half a centimetre, about a quarter of an inch) fossil they named *Ikaria warioonta*. Its tiny body shows signs of segmentation, with a definite head and tail, indicating the presence of a mouth, gut, and anus. These features strongly suggest that it, like *Spriggina*, was a bilaterian.

---

[5] Echinoderms are a big group of animals with fivefold symmetry that includes starfish, sea urchins, brittle stars, and sea cucumbers.

[6] The old name for the period we now call the Ediacaran was the Vendian, hence the name Vendobionts.

What makes *Ikaria* so special is that on the surface where it was found, it is closely associated with a distinctive trace fossil called *Helminthoidichnites*. The association between the two is so close that for once we can link a trace fossil with the animal that made it. It's almost certain that *Ikaria*, as it grazed its way through the microbial mat, created the *Helminthoidichnites* trace fossil. Although in the Flinders Ranges we cannot find *Ikaria* in sediments older than those containing other Ediacara fauna, we do find *Helminthoidichnites*. This suggests that *Ikaria*, or something very like it, existed at the time and has not been preserved, pushing the bilaterian record back to beyond 560 million years. I have marked the occurrence of *Ikaria* with a white star on Figure 3.2. Why is *Ikaria* so important? Well, *we* are bilaterians – so something like *Ikaria* is one of our very ancient ancestors.

The Flinders Ranges of South Australia are not the only place we can find Ediacaran fossils. Examples of the fauna have been documented from sites around the world. Specialists working on the Ediacarian fauna have grouped the fossils into three distinct assemblages. Most, but by no means all, palaeontologists believe these three biotas succeed one another in time. The oldest, which is between approximately 579 million and 570 million years old,[7] is the Avalon assemblage, named after the Avalon Peninsula in Newfoundland. It consists of a limited assemblage of fossils, mostly of frond-like organisms that lived in deep water. The Australian fossils belong in the second group and date to between 575 Ma and 550 Ma. Known as the White Sea assemblage after the region in Russia where they are also found, this is the most diverse group of Ediacaran fossils of all. Slow-moving grazers and filter feeders, this assemblage lived in shallow water on or in dense microbial mats that seemed to be ubiquitous at the time. The youngest group of Ediacaran fossil assemblages is found in Namibia: called the Nama assemblage, it ranges in age from 550 Ma right up to the base of the Cambrian at 541 Ma. Although the Nama

---

[7] Rocks of this age are difficult to date. This means that the dates in this and the following section are all approximate.

assemblage also lived in shallow water, it is distinctly different to the underlying White Sea assemblage. It's not very diverse: although it retains some of the animal groups found in the White Sea assemblage, most are missing. It does, however, yield more burrowing forms and even some species that had evolved hard parts.

THE CAMBRIAN EXPLOSION

If the pace of evolution has speeded up through the Cryogenian and Ediacaran periods, near the base of the Cambrian it goes into overdrive. Appearing in the latest Ediacaran is a group of fossils known as the small shelly fauna or SSF for short. The SSF live up to their name. They are indeed small, usually only a few millimetres in size, and they are the first record of animals with mineralised shells. They first appear only sporadically in the late Ediacaran, but both their numbers and diversity rapidly expand in the early Cambrian seas. The shells are phosphatic, and many of the better-preserved specimens are clearly early forms of today's shelly animals: the molluscs and brachiopods. But there are other, more enigmatic forms, and we don't know what animal group they should be assigned to. The SSF disappear in the early Cambrian (at about 510 Ma), probably because of changes in sea-water chemistry that resulted in the reduction of the amount of phosphate available to build their shells.

Alongside the SSF, the fossil record reflects changes in both the number and style of trace fossils. I have marked this as 'bioturbation' on Figure 3.2. Trace fossils from the Ediacaran assemblage are simple horizontal burrows and trails. However, as we approach the Cambrian, the number of trace fossils increases, and they become more complex. Some branch; others start to burrow vertically. Animals, almost certainly some form of worm, are churning through the sediment looking for food. This sudden increase of burrowing animals led the late Martin Brasier to christen this interval the 'carnival of worms'.[8] Early in the Cambrian, the trace fossils start to record a switch from grazing to scavenging, then finally to hunting. The

---

[8] In his book *Darwin's Lost World*.

appearance of predators reflects the fact that the planet's ecosystem has begun to become more complex.

The appearance of the SSF and the changes in bioturbation mark the beginning of an increase in the number and diversity of animals that has never been repeated in the history of life. In the 20 million years between the base of the Cambrian (540 Ma) and about 520 Ma, almost every invertebrate phylum that currently exists appears in the fossil record. For a brief period, the rate at which diversity increased soared. In 1997, Jack Sepkoski estimated that if diversification had continued at that same rate there would be 1×10^{60} families of multicellular animals present on Earth today. This huge number (1 followed by 60 zeros) vastly exceeds even the wildest of estimates of the number of species on the planet, let alone families! This spectacular increase in diversity is commonly referred to as the 'Cambrian Explosion', and in Figure 3.2 I have shown it as a vaguely triangular shape with its width representing the level of diversity: the wider the shape, the higher the diversity. The shape itself is purely diagrammatic; in reality, diversity did not increase smoothly but instead went through periods of rapid increase alternating with intervals of relative stability.

I am always nervous when we refer to this remarkable event as an 'explosion'. This is not because I want to downplay its significance: it was unprecedented, and nothing on its scale has occurred since. My concern is that the term could be misleading. From the perspective of deep time it was an 'explosion': it took 20 million years to evolve all those phyla, and that is almost a blink of a geological eye. However, from a human perspective it took a lot of time: it's not as if all the phyla just magically popped into being. But in the human time frame, that's what the word explosion implies – something almost instantaneous. Some creationists forget that 20 million years is still a very long time and emphasise the 'instantaneous' connotation, then suggest that the explosion is a result of special creation. Of course, as well as having to ignore the 20 million years it took to evolve all those animals, they have to wilfully ignore all the fossils (including my beloved Ediacaran forms) that we find *before* the base of the Cambrian.

The animals that evolved early in the Cambrian Explosion are well documented in four spectacular deposits. All four are referred to as *Lagerstätten*, a term applied to sites yielding exceptionally well preserved fossils. Three of these deposits range in age from about 522 Ma to 517 Ma. Although there is considerable overlap, in order of decreasing age they are the Chengjiang fauna from China, followed by faunas from Sirius Passet (Greenland) and Emu Bay (Australia). The most famous of the four Lagerstätten is the slightly younger (about 510 Ma) Burgess Shale from the Rocky Mountains of Canada. This deposit was the basis of the book by the late Stephen Jay Gould, *Wonderful Life: The Burgess Shale and the Nature of History.*

The fossils from these four Lagerstätten reveal a rich assemblage with a very diverse range of organisms, including arthropods, early trilobites, sponges, and worms. They also contain what are thought to be the earliest chordates, the group that gave rise to the vertebrates. In a clear sign of the development of complex ecosystems, the first real top-level predator evolves. It's called *Anomalocaris*, and it makes its oldest appearance at the Chengjiang site, although it is found in all the others. While the rest of the fauna was relatively small, *Anomalocaris* was a monster, reaching a metre or so in length. Some species of *Anomalocaris* gave up the predatory lifestyle and became filter feeders, using large, modified grasping arms to strain food from the water and pass it to their mouth.

I have been lucky enough to visit the Emu Bay site on Kangaroo Island in South Australia. While the other three sites are in spectacular or exciting locations such as high in the Canadian Rockies or the north coast of Greenland, the Emu Bay location is a most unprepossessing small hole in a field on a small island off the south coast of South Australia. But this small hole has produced wonders, including the beautifully preserved eyes of *Anomalocaris*, showing that they were, like insect eyes, compound – that is, composed of hundreds of tiny facets, each with a separate lens.

The fossil assemblages recovered from each of these Lagerstätten amply demonstrate that by the early Cambrian the first fully functioning complex 'modern' ecosystem had been established. It contrasts sharply with the much simpler Ediacaran ecosystem, which consisted largely of animals that fed either by grazing algal mats or by absorbing nutrients from sea water. For most of life's history, the biota was dominated by single-celled organisms. In the Ediacaran, we move from this simple world to a more complex one, where multicellular animals become common. Sediments from the early Cambrian record the appearance of hunters and prey, filter feeders, grazers, burrowers – in short, everything we would expect in a complex ecosystem, and that is exactly what is needed for the Earth System to become fully functional.

What caused the Cambrian Explosion? As with most of these ancient events, there have been many suggestions, and it is hard to prove any of them conclusively. Some suggest that what we are seeing is a burst of evolutionary experimentation following the appearance of mineralised hard parts. Another is that the trigger was a sudden injection of nutrients into the oceans. All around the world at about the same time as the start of the Cambrian Explosion (although there are some issues with the timing) is evidence for a massive period of erosion. The result of this erosion is most clearly seen in the United States' Grand Canyon where the 'Great Unconformity' – a vast period missing from the rock record – separates rocks of up to 1.8 billion years old from Cambrian rocks of a mere 525 million years old. The remains of all the weathering it took to produce that gap in deposition would be washed down rivers to the ocean where it would supply huge quantities of various nutrients, including phosphate, calcium, and silica in a synchronous global burst, possibly providing the fuel for the Cambrian Explosion.

There is a growing body of thought that suggests the explosion doesn't need much explanation at all. People supporting this idea believe that the event has its roots deep in the Ediacaran. This suggests that the transition to the Cambrian wasn't a wholesale replacement of the older Ediacara fauna. Instead, the transition reflects the appearance of successive faunal assemblages that evolved in response to the complex environmental changes that were happening at the time. I have some sympathy for this perspective – but I would still like to understand why the rate of diversification increased so suddenly at the beginning of the Cambrian. The answer almost certainly lies in the physical, chemical, and biological changes that were happening at the time.

I have spent a lot of time describing this interval of deep time. This was not simply because it is such an interesting period in the history of life, it is also the interval when the modern Earth System began. I believe that to deal with the environmental issues we are facing we must understand both how the Earth System operates now, and how it has operated through the long history of the Earth. To demonstrate how life and the biosphere essentially drove the appearance of the Earth System, I want to discuss one thread in the complex geochemical history of the Ediacaran–Cambrian transition: the amount of oxygen in the atmosphere. An examination of the deep history of oxygen will allow us to look at the way life and the physical environment interacted as the Earth System became established. Most multicellular life requires oxygen to live, and in general the more complex the organism the higher its oxygen demands. This means that it's likely that oxygen levels play a role in explaining not only the diversification across the Ediacaran/Cambrian boundary, but also why it took so long for multicellular life to appear in the history of the Earth.

THE DEEP HISTORY OF OXYGEN

For billions of years – in fact for most of the Earth's existence – life was limited to single-celled forms. Then, after a series of global glaciations, multicellular animals started to appear: first sponges, then the faunas of the Ediacaran Period, followed by the Cambrian Explosion. Until a few years ago, the reason behind this pattern of evolution seemed obvious: increasingly complex life followed increasing levels of oxygen in the atmosphere. However, the latest research into the complex mechanisms behind the accumulation of oxygen on Earth suggests that this story is far too simple.

When life first evolved on Earth, the atmosphere contained either very little free oxygen or possibly none at all. Under these conditions, the evolution of multicellular life was almost certainly impossible. However, some of the oldest fossils ever found have been identified as a type of single-celled prokaryote called cyanobacteria. These relatively simple organisms evolved a whole new way to survive – photosynthesis. The standard story, which I repeated to my undergraduate classes,[9] claimed that once these tiny organisms started to produce oxygen it slowly accumulated in the water; then about 2.45 billion years ago it started to accumulate in the atmosphere. Here we see the first stirrings of the Earth System – cyanobacteria to harness the Sun's energy to produce the sugars needed to support life, and a by-product of this process is oxygen. Life is starting to change the atmosphere. Initially the amount of oxygen was well below present atmospheric levels, but as more and more cyanobacteria appeared, oxygen levels started to rise, although it was thought that they did not reach today's levels until well into the Paleozoic. However, it was supposed that from about the time of the Snowball Earth glaciations onwards, enough oxygen had accumulated to permit the evolution of complex multicellular animals. Well, that was the story, but it's now become clear that it is wrong.

There is no doubt that the evolution of complex animals is linked with the amount of available oxygen, but far from passively evolving into an environment with an appropriate level of oxygen, these early animals appear to have actively contributed to the creation of that environment. It's a far more interesting story, and it touches on the very beginning of the Earth System. To understand how that happened, we need to know a little more about what controls the level of oxygen in the atmosphere. I am only attempting a brief outline

---

[9] But no longer.

here – the details are extremely complex, and not always agreed on by scientists who research the history of the early Earth.[10] It may sound odd, but if we want to begin to understand what controls the level of oxygen on Earth, we should start by looking at some very old iron ore.

Today, much of Australia's wealth is generated through its exports of iron ore. Most of this ore is derived from a particularly iron-rich rock called a banded iron formation (BIF). BIFs are very common around the world, and they are particularly abundant in Western Australia. They are spectacular rocks to look at. The Melbourne Museum has a huge, polished specimen on display. Fine layers of a rich dark red, gold, and grey swirl around like the marbled end paper in an old book. The red layers are jasper, a form of silica; the golden bands are another form of silica called tiger-eye (you may have seen this form used in jewellery). What the miners are interested in is the dark grey layers, which are iron oxide.

BIFs are most common in rocks older than 2.45 Ga.[11] They were formed at the bottom of an ocean that was very rich in a form of iron that can be dissolved in water. The iron itself was derived from the weathering of continental rocks and transported to the ancient ocean dissolved in rivers and streams. Today's ocean waters contain almost no dissolved iron – although iron-rich sediments are still being eroded from continental rocks by rivers, and river water eventually reaches the ocean. This is because the level of oxygen in our present atmosphere is so high that it quickly oxidises the dissolved iron in the river water. The product of this reaction is iron oxide, which is essentially rust. Iron oxide is insoluble, so it precipitates out of solution before it reaches the sea. Billions of years ago, when BIFs were forming, there was no oxygen in the atmosphere, and all the iron weathering out of rocks could be transported to the ocean, making it very rich in dissolved iron.

---

[10] If you want to understand more about the details of what happened, you should read Donald Canfield's excellent book *Oxygen: A Four Billion Year History*.

[11] Ga = giga annum, 1 billion years.

To produce BIFs, something is needed to make the iron precipitate out of the ocean water and be deposited on the ocean floor: we need oxygen. We have seen that well before 2.5 Ga, oxygen-producing cyanobacteria were living in the iron-rich ocean – after all, fossils of cyanobacteria have been found in rocks over 3 billion years old. It's easy to imagine that in some parts of the ocean (say, around the area that would eventually become Western Australia) the oxygen produced by the cyanobacteria would rise, triggering the reaction that precipitates iron oxide out of solution. This results in the deposition of one of those grey iron oxide bands in the Melbourne specimen. The production of iron oxide continues until all the oxygen is consumed. When this happens, the precipitation of iron oxide stops, and normal sedimentation resumes. This normal sedimentation will then continue until the level of oxygen again reaches a critical level, and the whole cycle will repeat.[12] I find it a fascinating idea that a significant part of the wealth of Australia is due to the actions of bacteria over 2 billion years ago.

The upshot of all this is that the presence of BIFs is an indication that at the time there was no oxygen present in the atmosphere, but that it was episodically present in the ocean. But at about 2.45 Ga, the number of BIFs found in the rock record reduces significantly, and this is almost certainly because from that point on low levels of oxygen were present in the atmosphere, stopping the transport of iron to the ocean. The point in time when the accumulation of significant quantities of oxygen in the atmosphere started is known as the Great Oxidation Event or GOE. As I noted above, there are some BIFs in sediments younger than the GOE (particularly in North America), and they are an important part of the story: I'll come back to them.

Complicating things a little are indications of early 'whiffs'[13] of atmospheric oxygen in sediments older than the GOE. The first record

---

[12] There have been other suggestions made about what triggers this precipitation of iron, such as ultraviolet light interacting with the surface waters. Ultraviolet would have been more intense at the time the BIFs were forming as there would have been no ozone layer to protect the planet's surface so it could have done the job, but varying levels of oxygen in the ocean are considered to be the most likely cause.

[13] This great phrase is not mine! It comes from the authors of the study that recognised this important event.

of a brief whiff of oxygen came from geochemists examining marine sediments dated between 2.6–2.5 billion years old. They recorded a temporary rise in the amount of the molybdenum in these marine sediments. Like iron, molybdenum is released as sediments are weathered. However, unlike iron, it can only be transported to the ocean if oxygen is present in the atmosphere. The increase in molybdenum in sediments of this age is strongly suggestive of a brief period when there was oxygen in the atmosphere, well before the GOE. Since the original research, several other whiffs have been documented: some reach back as far as 3 billion years. Why only passing whiffs? Why, once oxygen started to accumulate in the atmosphere, didn't it just keep on building up, ushering in complex animals earlier? To answer these questions, we need to understand the basic controls on oxygen accumulation in the atmosphere.

The primary control of the level of free oxygen in the atmosphere is the burial of carbon. The first time I read that, my brain went into meltdown, and I had to go away and have a nice cup of tea. What on Earth does the burial of carbon have to do with the accumulation of oxygen? It took me a while to get my head around exactly what it means. Cyanobacteria produce the oxygen – but most of the oxygen that they produced is reused when they die and their remains decompose. During decomposition, the carbon in their bodies combines with oxygen they made when they were alive, producing carbon dioxide and leaving just a trace of oxygen. The only way to increase the level of free oxygen significantly is to bury the bodies away in sediments before any available oxygen reacts with the carbon they contain – in effect, bury the carbon. The more unreacted carbon that is buried, the larger the net supply of oxygen to the ocean, and once the oceans become saturated with all the oxygen they can contain, its level in the atmosphere will eventually start to rise. Increasing the numbers of oxygen-producing cyanobacteria without increasing the removal of carbon will not result in any significant increase in atmospheric oxygen. Carbon is buried all the time; it remains as organic-rich rocks such as coal. Today, when we burn coal, we are effectively completing the decomposition process that was started millions of years ago, combining modern atmospheric oxygen and the ancient buried carbon, and releasing carbon dioxide.

The whiff of oxygen at 2.5 Ga is accompanied by geochemical evidence of a significant increase in the amount of carbon being buried. The reason for a sudden increase in carbon burial is unclear, but it may be a result of an increase in nutrient supply to the oceans which allowed organisms to bloom. But the question we are addressing is, once it reached the atmosphere, why didn't oxygen just go on accumulating? The answer is volcanoes. If the accumulation of oxygen in the atmosphere is a result of burying carbon, then its removal is a result of oxygen combining with volcanic gases. Oxygen is a highly reactive gas, and once in the atmosphere it will quickly react with gases produced during eruptions. Particularly effective at removing oxygen is hydrogen, a very common volcanic gas that swiftly joins with oxygen to produce water (H_2O).

To sum up, at its simplest the level of oxygen in the atmosphere can be considered to be a balance between the burial of organic carbon and the production of volcanic gases. Each whiff of oxygen was a time of increased carbon burial that allowed oxygen to accumulate in the atmosphere. The oxygen stayed in the atmosphere long enough to produce the molybdenum peak in the ocean. It could do this because the amount of volcanic activity was low. The level of volcanic activity has risen and fallen through deep time, depending ultimately on churning in the mantle deep below the Earth's crust. So, after a brief lull, the level of volcanic activity increased, causing the level of oxygen in the atmosphere to fall again to near zero. The transport of molybdenum was stopped but the transport of iron was resumed. Once in the ocean, the iron was available to be deposited as BIFs.

The GOE is obviously different. At that point, things changed more or less permanently: significant levels of oxygen accumulated in the atmosphere. The lower numbers of BIFs in younger sediments means that from 2.45 Ga onwards there was always some oxygen in the atmosphere. The reason that oxygen doesn't disappear completely after the GOE is possibly due to a permanently lowered level of volcanic activity. When the planet first formed, it was very hot, and the volcanic activity was high. Over billions of years it has cooled, decreasing volcanic activity. By the time we reach the GOE, the level of volcanic activity was so low there wasn't enough hydrogen being produced to strip all the oxygen out of the atmosphere.

But things are never that simple, and as I have noted, the record of oxygen is particularly complex. Remember the few BIFs that are younger than the GOE? Their presence, together with some detailed geochemical analyses (including molybdenum analyses), suggests that following the GOE, the level of atmospheric oxygen didn't go on increasing, finally reaching a level that would allow the evolution of complex animals. In fact, it fell. Detailed geochemical analysis suggests that following the GOE, oxygen levels fell to perhaps as low as 0.01% of present-day levels. The reason for this drop in atmospheric oxygen is uncertain, but it may have something to do with the huge amount of organic carbon that had been buried during the GOE itself. If these carbon-rich sediments were brought back to the planet's surface as part of the natural rock cycle, weathering would expose them to the atmospheric oxygen. The carbon and oxygen would then react, yielding carbon dioxide and completing the cycle of decomposition.

Let's concentrate now on the interval of time between 700 million and 440 million years ago, shown in Figure 3.2. Down the right-hand side of that figure, I have shown, in its simplest form, oxygen levels in the ocean. Ignoring for the time being the light grey bars, between the base of the column about 700 million years ago and about 580 million years ago (just after Gaskiers Glaciation) I've shaded the column black. This represents a time where, while the waters at the very surface of the ocean may have been oxygenated, most of the ocean was anoxic (lacking oxygen). The only multicellular animals alive at this time were sponges. This isn't a surprise, as work on living sponges suggests that they can survive quite nicely at very low oxygen levels. From about the 580 million years mark (just following the Gaskiers Glaciation), there is a significant change in the oxygen levels in the deeper parts of the oceans. From this point onwards, the oceans became environmentally heterogeneous, with some regions becoming oxygenated whereas others remained anoxic. This is reflected in the diagram by a change in shade to dark grey.

The change in the distribution of oxygen that affected the deeper parts of the ocean was not reflected in the surface waters. This is because oxygen was present in the atmosphere throughout this time interval and the gas could be absorbed by water at the ocean's surface, ensuring that a shallow layer of oxygenated water was continuously present. From the early Cambrian (around 480 million years ago), presumably in response to increasing levels of oxygen in the atmosphere, the oxygenated surface layer started to expand. I have marked this expansion with an arrow to the right of the oxygen history column in Figure 3.2. The result of this expansion was an increase in the levels of oxygen in much of the water column. However, the oxygenated waters never reached the deepest parts of the ocean, which remained anoxic. Nevertheless, there is no doubt that by the early Cambrian a large proportion of the world's ocean waters contained significant levels of oxygen, but the level was still well below what we find in today's ocean.

The shift from an anoxic ocean to a heterogeneous one that followed the Gaskiers Glaciation coincides with the single-celled acritarchs becoming increasingly common and diverse. As they diversified, they also got larger. Some of the biggest acritarchs that ever existed (over 150 micrometres – today they average about 20) are found at this level. When they died, because of their size these large forms would have settled to the sea floor faster than their earlier, smaller ancestors. This allowed less time for decomposition and promoted the liberation of oxygen. In other words, the increase in oxygen was driven by an increase in the number of acritarchs and the evolution of larger forms.

The light grey bars across the oxygen column represent transitory increases in the oxygen level of almost the entire ocean. These represent times when oxygenated surface waters rapidly expanded, and even the deeper parts of the ocean became well oxygenated – they have been referred to as oceanic oxygenation events or OOEs. The one that immediately follows the Gaskiers Glaciation is coincident with significant changes in the biosphere. It's probably no coincidence that the Avalon Ediacaran assemblage evolves at about this time.

The appearance of the larger multicellular Ediacaran organisms would have resulted in even more carbon being transported to the sea floor. It was not just a result of their bodies getting larger, but because they were also getting more complex. For the first time, multicellular animals were producing faecal pellets, adding to the carbon being supplied to the ocean floor and resulting in increasing levels of oxygen.

So efficient was the burial of carbon and the liberation of oxygen that by the time of the White Sea assemblage, the amount of oxygen in the atmosphere may have reached about 15% of its current level, resulting in another OOE. In this case, even the deepest parts of the ocean may have been oxygenated. So again it's no surprise that the White Sea assemblage is the most diverse of the three Ediacaran biotas. This shows how wrong was the old story I told my students. The first multicellular animals didn't evolve into a nicely oxygenated environment that was waiting to receive them. It was their evolution that was responsible for the accumulation of oxygen.

But it didn't last. Soon after the White Sea assemblage, the oceans lapsed back into near anoxia. The result was the Nama assemblage, a low-diversity group of fossils consisting largely of survivors of the White Sea assemblage. None of these survivor species are found in sediments from the overlying Cambrian Period. The Nama assemblage does, however, include several new species, some with hard parts, and a few of these continue into the younger sediments. The reason for the late Ediacaran reduction in oxygen is again not well understood, but it may have something to do with the evolution of animals that could burrow vertically into the sea floor. Rocks containing the Nama assemblage do record an increase in the amount and style of trace fossils activity, which increases as we move into the

Cambrian. We have talked about this increase in trace fossils earlier, and I've marked it as bioturbation on Figure 3.2. I'll come back a little later to the idea that there is a relationship between lowering oxygen levels and bioturbation.

Because most of the Ediacaran fossils go extinct around the beginning of the Cambrian, some palaeontologists claim that the base of the Cambrian represents the Earth's first mass extinction. But I have already said that my sympathies lie with the idea that the Ediacaran–Cambrian transition is one of changing faunas rather than a sudden replacement. What it looks like to me is that a biota, stressed by low oxygen, was simply replaced, over a period of about 10 million years, by a new assemblage of animals that could survive in the low-oxygen conditions. The descendants of this assemblage will come to fill the oceans in a massive burst of diversity: the Cambrian Explosion.

The early Cambrian OOE event marked on Figure 3.2 coincides with this extraordinary event, and it is easy to understand why. In the same way that the arrival of the Ediacara fauna drove the earlier accumulation of oxygen in the ocean, the increase in the number and diversity of complex organisms that is the Cambrian Explosion triggered a significant increase in oxygen levels. The Cambrian oceans must have been richer in oxygen than the Ediacaran oceans: they could support large numbers of very complex organisms. However, despite this, oxygen levels in the Cambrian seas were still significantly less than we find today.

But even these levels of oxygen couldn't last, and they dropped back: there are even a couple of intervals when deep anoxic waters almost reached the surface (the opposite to an OOE). One suggestion made to explain why the Cambrian oceans moved back to a lower oxygen mode is that it was a result of the change in the record of bioturbation I talked about earlier in this chapter. As both the level and style of bioturbation increased, it may have altered the geochemistry at the interface between the ocean waters and the sea floor. This in turn altered several of the Earth System's feedback loops that cycle important elements through the biosphere, which ultimately reduced the oxygen level. If that's the case for the early Cambrian anoxia, perhaps the same is true for the late Ediacaran oceans. Perhaps the anoxia that caused the transition to the lower-diversity Nama assemblages was a result of animals of the White Sea assemblage evolving the ability to burrow.

THE EARTH SYSTEM EMERGES

The accumulation of oxygen in the atmosphere and the oxygenation of the ocean that occurred over the billions of years prior to the Cambrian demonstrates a complex interplay between life and the physical environment. In this interplay we see the earliest stages of the evolution of the Earth System. But this early Earth System was not very good at maintaining climatic conditions. On at least three occasions over this interval, the Earth was plunged into glaciations that froze the planet from the poles to the Equator. We couldn't by any stretch of the imagination call that a stable climate.

Two prerequisites were needed for today's Earth System to become fully operational. Firstly, we needed animals – large multicellular animals – and they started to appear in the Ediacaran but really became diverse during the Cambrian Explosion. Secondly, we required a complex diverse ecosystem – and we see that happening in the early part of the Cambrian. The events that took place across the Ediacaran–Cambrian transition mark the emergence of the modern Earth System, bringing with it a level of relative climatic stability. We shouldn't be surprised at the close association between the evolution of large multicellular animals and the Earth System because, as palaeontologist Nicholas Butterfield says, 'they invented it'.

4 Documenting Ancient Biodiversity

INTRODUCTION

Mass extinctions are one of the hazards of living on planet Earth, and I am going to argue that they have played an important role in the history of life. However, if we are to understand that role, we need to be able to document the story of the planet's biodiversity. Historians telling the story of human civilisations are able to marshal an enormous amount of detail, from the rise and fall of empires down to what the everyday people had for lunch. These human histories take place over relatively limited time spans of hundreds to perhaps thousands of years. Palaeontologists, on the other hand, telling the story of ancient biodiversity, lack the level of detail that historians can muster, but they more than make up for it in the length of time involved. Palaeontologists are attempting to document the diversity of life over the *billions* of years of deep time.

In this chapter I will review just a few attempts at documenting ancient biodiversity. As usual, the limitations of the fossil record mean that we will be looking largely at marine fossils over the past 500 million years or so. As we will see, not all palaeontologists' estimates of the amount of ancient biodiversity agree with each other. Given what is being attempted, that's hardly surprising. They do, however, all agree on three key points. Firstly, there are more species alive on the planet today than at any other time in its history. Secondly, the accumulation of biodiversity from a single species to the approximately 8.7 million we see today has been anything but a steady increase: there have been periods of great growth in biodiversity like the Cambrian Explosion; and at least five catastrophic declines, each associated with a mass extinction (Figure 2.3). And finally, it is clear that mass extinctions, as well as exerting some level of control over the level of biodiversity, can significantly alter the composition of the planet's biota.

As well as documenting ancient levels of biodiversity, the studies I will outline here also demonstrate the increasing sophistication of the techniques that palaeontologists are using to interrogate the fossil record. Over time, alongside the development of the modern Geological Time Scale, improved databases and more powerful statistical analyses have enabled palaeontologists to significantly improve their estimates of biodiversity, perhaps edging us closer to some level of agreement.

FIRST ATTEMPTS

I'm going to start with what is possibly the earliest attempt at documenting changing biodiversity through time. It was carried out in 1860 by the English geologist John Phillips.[1] This early attempt at documenting diversity is shown in Figure 4.1.

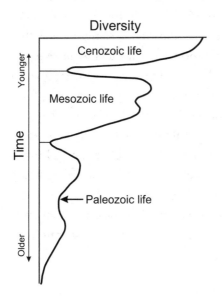

FIGURE 4.1 Diversity of Phanerozoic life as estimated by John Phillips in 1860.

---

[1] Phillips was the nephew (by marriage) of the more famous English geologist William Smith, star of Simon Winchester's book *The Map That Changed the World*.

Although Phillips did not have the benefit of a modern Geological Time Scale, and he certainly didn't think in terms of millions of years, he could place the rocks he was studying in chronological order. On the vertical axis of Figure 4.1, the oldest rocks are at the base and the youngest at the top. He assigned the rocks to three successive intervals of time – the Paleozoic, Mesozoic, and Cenozoic.[2] What I have labelled on the horizontal axis as 'diversity', Phillips called 'the variety of life'.

Phillips started the construction of his diversity curve by simply counting the number of fossil species found in each of the three intervals of time that he had recognised. His raw data suggested that rocks from the Paleozoic contained far more fossil species than those assigned to the Mesozoic, and that Cenozoic rocks contained the lowest number of fossils of all. But Phillips recognised that there was a bias built into this data. His study was restricted to rocks from Great Britain, and because of British geology, the thickest sequence of rocks is Paleozoic in age and the thinnest are from the Cenozoic. This meant he had more opportunity to collect fossils from Paleozoic rocks than from the younger time intervals. It created a sampling bias that made it seem as if there were more species in the older rocks than the younger. To overcome this, he did something quite smart. He divided his species count by an estimate of the thickness of the rocks from each of the three time intervals he had recognised and came up with a measure of the number of species per 1,000 feet of rock. In doing so he removed, or at least reduced, the bias that his collecting had introduced. So, although he did not provide a horizontal scale for the figure above, it's effectively the number of species per 1,000 feet of rock. Thus, as well as producing perhaps the first history of ancient biodiversity, Phillips took the first tentative steps towards minimising the biases inherent in the fossil record.

[2] The names have Greek origins. Paleozoic = ancient life; Mesozoic = middle life; and Cenozoic = recent life.

Let's take a closer look at the plot in Figure 4.1: Phillips recognised three distinctly different, consecutive assemblages of fossils that he named Paleozoic life, Mesozoic life, and Cenozoic life. Each successive 'life' is more diverse than the preceding one, and they are separated, one from another, by a 'bottleneck' of lower diversity. Today we know that those bottlenecks represent mass extinctions. The boundary between Phillips' Paleozoic life and Mesozoic life is the end-Permian mass extinction, and the boundary between Mesozoic life and Cenozoic life is the end-Cretaceous mass extinction.

So, what does Phillips' curve tell us about overall biodiversity of ancient marine life in Britain? From the beginning of the Paleozoic, diversity increases fairly steadily through that period. Following a major fall, it rises again during the Mesozoic, but the biota is significantly different from the Paleozoic. The level of biodiversity of this new life grows at an even faster rate than during the Paleozoic. But this increase in biodiversity is stopped by another, even larger, decline, and there is another transition to a new life, the Cenozoic.

As early as 1860, Phillips recognised that large-scale crashes in biodiversity triggered major transitions in the make-up of the planet's biota. Phillips did not have either the detailed understanding of the geological time scale or access to the better dating of rocks we have available today, so he shows the two drops in biodiversity as being somewhat gradual affairs. Today we recognise that they were, even on a human time scale, rapid. Some estimates suggest that it took a mere 20 thousand years to decimate the planet's biodiversity at the end of the Permian.

SEPKOSKI 1984 – A CLASSIC STUDY

Modern studies have, of course, provided us with a far greater level of detail and precision. In 1984, 125 years after John Phillips published his diversity curve, Jack Sepkoski published one of the first modern attempts at documenting Paleozoic diversity. Sepkoski's study is a classic that is well worth spending some time on.

For his study, Sepkoski used the same data set that had allowed him and Raup to recognise the Big Five mass extinctions (Figure 2.3). Because of the difficulty in counting species in the fossil record, Sepkoski's study used a higher taxonomic level: he focused on the level of *family*. But before he could start his analysis, Sepkoski had to address a statistical problem called the *Pull of the Recent*. This is an effect that results in an artificial increase in diversity as we approach the present day, making it look as if there are very many more species in younger rocks than in older.

Several factors contribute to the Pull of the Recent. The first is similar to the problem John Phillips faced: the amount of rock available to sample can influence any count being made. Globally, there are more outcrops of younger rocks than older ones, so it's likely that we will collect more fossils from them. This is because over time, the older a rock is, the more likely it is to become 'recycled' – eroded, or deeply buried, or pushed down into the Earth's mantle at plate boundaries. There are simply more younger rocks at the Earth's surface than older ones.

Secondly, the composition of the rocks being sampled can also make a difference: it is easier to collect fossils from younger, softer rocks than older, harder ones (older rocks are more likely to have been deeply buried, metamorphosed, and deformed). Together, these effects could artificially raise the recorded level of diversity in younger sediments. Because of the way Sepkoski's data set was structured, there was little he could do to minimise these effects.

However, he could do something about the third and possibly most significant contributing factor – families that have representatives living today. Including these species in a diversity count will lead to an overrepresentation of living taxa when compared with species that have gone extinct prior to the recent. I have attempted to explain how this happens in Figure 4.2. Down the left-hand side is the same geological section we saw in Figure 2.2. This time, however, I have divided it up into five time intervals labelled T1 (the oldest) to T5 (the youngest). The recent – today – is sitting at the top of T5.

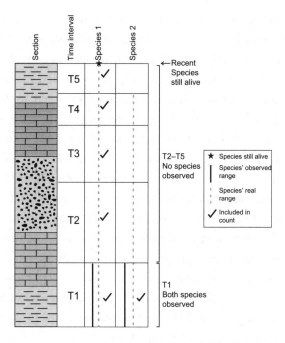

FIGURE 4.2 A model of the 'Pull of the Recent'. The vertical axis is the same geological section as Figure 2.2. Plotted alongside are five time intervals (T1–T5) and the ranges of two imaginary species – see text for details.

Plotted alongside are the ranges of two fossil species. The solid black lines represent the *observed* range of each species. Both species are observed in T1, then they disappear from this section. For reasons explained in Chapter 2, the disappearance may not represent the extinction of these species – it may be a local disappearance while elsewhere they are living happily. I have shown this with the grey dashed lines that represent the *true* range of the species. In the case shown here, species 1 is still alive today, so we know that its true range is T1 to T5 – even though we don't record its presence in the section. Species 2, on the other hand, is not alive today, and although it has a true range of T1 to T4, we only record it in T1. We cannot include the rest of its range because, based on the data we have available, the species appears to have gone extinct at the end of T1.

The problem comes when we try to count the number of species in each time interval in order to estimate its biodiversity. Because species 1 is observed in T1 *and* the recent, we can be certain that it should be present in all the time zones in between, and it would be counted in each (the ticks in Figure 4.2). However, when we look at species 2 – only observed in T1, even though it really ranges into younger intervals – in this section these appearances aren't recorded, so we only count its appearance in T1. This means that for time intervals T2 to T4, we are recording the presence of species 1 but not species 2, and we are doing this simply because species 1 is still present today. This pattern, if repeated in a section with many species, would lead us to significantly overestimate the contribution of species alive today compared with those that went extinct prior to the recent. This bias is the major contributor to the Pull of the Recent. To overcome this, Sepkoski filtered his data by simply excluding any families that had any species living today. We will come back shortly to talk about how significant the Pull of the Recent is.

Sepkoski's estimate of total global marine biodiversity is reproduced in the 'spindle diagram' of Figure 4.3. The vertical axis is time; the number of families is on the horizontal axis. The width of the spindle represents the estimated number of families over time. Remember, each of these families probably contains many species. The horizontal dashed lines mark the presence of the 'Big Five' mass extinctions.

The Sepkoski curve suggests that biodiversity starts from a very low level in the early Cambrian. It rises steadily, reaching about 150 families by the end of the Cambrian. The next youngest period, the Ordovician, is marked by an extremely rapid rise in diversity, reaching around 500 families by its close. I have marked this as the 'Ordovician expansion' in the key events column in Figure 4.3. It has been suggested that this rapid increase in diversity reflects a significant increase in the concentration of oxygen in the atmosphere. Regardless of its cause, this global increase in biodiversity terminates with the end-Ordovician mass extinction, the first of the Big Five

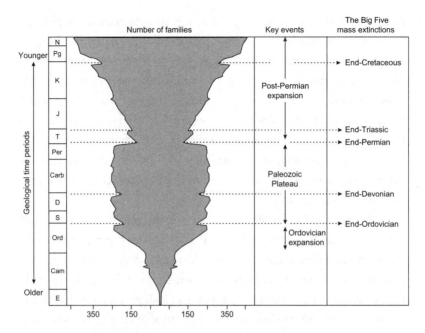

FIGURE 4.3 Sepkoski's estimate of biodiversity over the Paleozoic. Geological periods are listed down the vertical axis, number of families along the horizontal. Key events in the history of biodiversity are shown, and the Big Five mass extinctions are marked by the dashed lines. Adapted with permission from Sepkoski (1984).

mass extinctions. This event is marked by a sharp drop in total biodiversity followed by an almost equally sharp recovery.

After the end-Ordovician mass extinction, biodiversity starts to rise slowly during the Silurian and Devonian, but is knocked back again by a mass extinction – the second of the Big Five: the end-Devonian event. Following this event, global biodiversity stays more or less flat until the end of the Permian. This interval from the end of the Ordovician to the end of the Permian represents about 200 million years of relatively stable levels of diversity and is often referred to as the Paleozoic Plateau (key events column, Figure 4.3).

The Paleozoic Plateau ends with an almighty crash, the end-Permian mass extinction – the biggest mass extinction of them all.

Biodiversity falls spectacularly – and it's slow to recover. Sepkoski's analysis dramatically shows how this event decimates the planet's marine life. He records an overall fall in diversity from around 500 families prior to the event to significantly less than 200 afterwards. And remember these are *families*, the vast majority of which contain many species. Looking at Figure 4.3, it's clear why the end-Permian event is known as the 'mother of all mass extinctions'.[3] The rise in biodiversity that starts immediately following the end-Permian extinction is cut short by the fourth of the Big Five mass extinctions: the end-Triassic event.

Following the double whammies of the end-Permian and end-Triassic events, biodiversity takes off and expands rapidly. This explosion of biodiversity continues right through to the present day. Even the end-Cretaceous mass extinction, which we will see is one of the largest of the mass extinctions, appears as a slight interruption in this continuous rise in biodiversity. I've labelled it as the post-Permian expansion on Figure 4.3. This spectacular, long-term rise in biodiversity is perhaps the most startling feature of Sepkoski's estimate of ancient biodiversity. It is certainly the most controversial.

The controversy surrounding the post-Permian expansion can be summed up in a single question – is it real? Some workers believe that much, if not all, of this increase in biodiversity is due to the *Pull of the Recent*, suggesting that the post-Permian expansion is merely a statistical artefact. They think that although Sepkoski filtered his database, he didn't do enough. I'll come back to this discussion towards the end of this chapter. Nevertheless, Sepkoski's work plainly shows the important role that mass extinctions play in influencing the level of biodiversity: time after time, diversity rises only to be knocked back by one of these events.

In his study, Sepkoski was interested in more than simply recording changing biodiversity through time. He also wanted to drill

---

[3] I don't know who first used this term, but it's accurate.

down into his biodiversity data and document any changes that took place in the make-up of the planet's marine biota over the same period. To do this, he filtered his database further. Firstly, he removed any animal families that had a poor fossil record (jellyfish, for example, which have no hard parts and are therefore difficult to preserve as fossils), and then he removed any families with doubtful taxonomic placement (notably the Cambrian *Archaeocyatha* I mentioned in the Brachina Creek field trip report). Once this filtering was complete, he ran a detailed statistical analysis of his database. This analysis involved looking for families of marine organisms that consistently grouped together. Essentially, he was looking, in a statistically meaningful way, for groups of organisms similar to Phillips' three 'lives'. Sepkoski's analysis suggested that there were in fact three such groups underlying the biodiversity signal: he called them evolutionary faunas. Unlike Phillips' 'lives', which were separate and sequential, Sepkoski's three evolutionary faunas overlap and interact with each other. In detail, Sepkoski's three evolutionary faunas are:

(1) *The Cambrian fauna.*

The Cambrian fauna has its origin in the late Ediacaran and is composed mainly of trilobites (which, after dinosaurs, must be one of the best-known of all fossils), together with elements of a small shelly fauna – notably hyolithids (a small enigmatic animal with a cone-shaped shell), simple molluscs, and representatives of the Brachiopods (a group that look superficially like clams but are actually unrelated).

(2) *The Paleozoic fauna.*

Members of this fauna include various corals, more advanced forms of Brachiopods, and Echinoderms (a group that includes sea urchins, starfish, and sea lilies). Trilobites continue to be an important component of the fauna. If I had to nominate a fossil icon for this fauna, it would be the ammonites. These animals lived in chambered shells and look superficially like the modern-day coiled-shelled nautilus. During the Paleozoic, ammonites were very diverse, and their lovely coiled shells were often twisted into bizarre shapes. Fossil ammonites are very common and highly decorative; I have a beautiful set of cufflinks made from an ammonite shell that I bought on a palaeontological pilgrimage to the English seaside resort of Lyme Regis.[4]

(3) *The Modern fauna.*

Despite its name, this fauna had its beginnings in the Cambrian but ends up dominating the marine faunas from the Triassic onwards. Important components of this fauna include Gastropods (snails) and Bivalves (clams and their allies), burrowing Echinoderms (sand dollars), Fish, marine mammals, and marine reptiles. Also significant is a group that includes crabs, lobsters, and shrimps – the Malacostraca. This fauna also records the growing importance of tiny animals called foraminifera. These single-celled creatures are related to the amoeba but secrete beautiful calcium carbonate shells.

Sepkoski's recognition of the three evolutionary faunas enables us to see what is underlying the ups and downs of biodiversity. Figure 4.4 sets out in detail the way these faunas interact through time. Again, I have presented them in the form of spindle diagrams, so the width of each spindle equates to the level of biodiversity within each group. The vertical axis shows the geological periods and the dashed lines mark the position of the Big Five mass extinctions.

As you might expect, the initial Cambrian rise in diversity is due to the developing Cambrian fauna. But this fauna runs out of steam by the middle of the Cambrian and slips into a slow decline. It undergoes a sharp fall during the end-Ordovician mass extinction from which it never fully recovers. Its final significant appearance is just before the end-Devonian mass extinction, although some elements of the fauna trickle on into younger time periods.

The rapid Ordovician expansion is all about the Paleozoic fauna: it simply explodes, quickly replacing the fading Cambrian fauna as the dominant group. Because it is so dominant, the fortunes of the Paleozoic fauna mirror the overall level in biodiversity. The rapid

---

[4] Lyme Regis, in Dorset, England, is a place inextricably linked with Mary Anning, a cabinetmaker's daughter who during the early 1800s supplemented her family's meagre income by extracting spectacular vertebrate fossils from the cliffs along the shore and selling them to wealthy collectors. It is also a place I'd happily retire to.

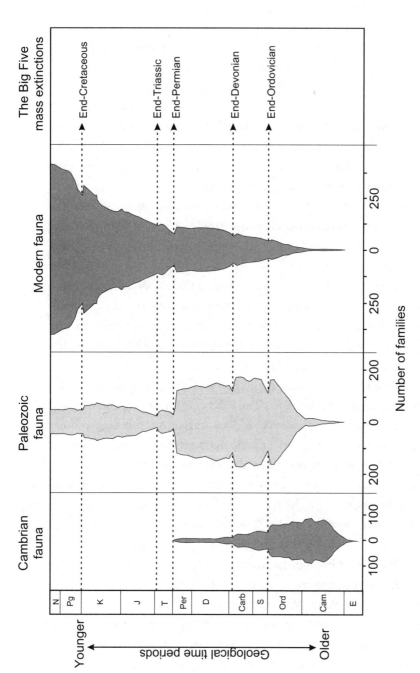

FIGURE 4.4 Sepkoski's three evolutionary faunas. Geological periods are along the vertical axis, the numbers of families are along the horizontal. The dashed lines mark the Big Five mass extinctions. Adapted with permission from Sepkoski (1984).

increase in diversity of the Paleozoic fauna is halted by the end-Ordovician mass extinction. Following this event, there is a slight rise in its diversity, but this is terminated by the end-Devonian event. Between the end-Devonian and end-Permian events, the biodiversity of Paleozoic fauna undergoes a slow but noticeable decline.

During the early Ordovician, as the Paleozoic fauna undergoes its meteoric rise in biodiversity, the Modern fauna begins a slow and steady climb. The end-Ordovician and end-Devonian events barely trouble the Modern fauna. It recovers quickly from each event, then continues its slow and steady increase in biodiversity. The increase in the Modern fauna between the end-Ordovician and end-Permian events almost exactly matches the slow decline in the Paleozoic fauna. We have already seen that the end-Ordovician and end-Devonian mass extinctions also play a significant role in stabilising biodiversity over this interval: every time that biodiversity starts to rise, it is knocked back by a mass extinction. But the slow replacement of the Paleozoic fauna with the Modern fauna is also in part responsible for the maintenance of the steady level of biodiversity that characterises the Paleozoic Plateau.

The end-Permian mass extinction profoundly changes the make-up of the planet's marine biota. The Paleozoic fauna is decimated, and it never recovers its abundance. It does stagger on, suffering another decline as a result of the end-Triassic mass extinction. Its biodiversity rises a little during the Jurassic and Cretaceous but this is stopped by the end-Cretaceous event. However, the Paleozoic fauna never completely disappears from the record. Sepkoski's data suggest that there are about 100 families of the Paleozoic fauna, including corals and brachiopods, remaining as we approach the present day.

The end-Permian event also results in the Modern fauna losing a significant amount of its diversity, but it quickly recovers. The rapid expansion of the Modern fauna that follows the end-Permian event is almost entirely responsible for the post-Permian expansion that characterises Sepkoski's diversity curve. It suffers substantial setbacks in the end-Triassic and end-Cretaceous mass extinctions but, as significant as they are, they result in only a temporary slowing of the relentless rise in its diversity.

To see the role of mass extinctions in the changing composition of the planet's biota, try to imagine what Figure 4.4 would look like without them. We can think of many possible histories. Could the Cambrian fauna have been a significant part of today's biota, if it had not been for the first three big mass extinctions? Would the Paleozoic fauna have kept on expanding, if it had not been for the end-Ordovician and end-Devonian events, and how would this have affected the Modern fauna? If the end-Permian event had never happened, would the Paleozoic fauna have exploded in the same way as the Modern fauna and become the dominant group in today's biota? Sepkoski's pioneering work showed us that mass extinctions are important and that they play a major role in controlling both the level of the planet's biodiversity and the composition of its biota.

BETTER DATA SETS, MORE UNCERTAINTY?

Since Sepkoski's classic 1984 paper, palaeontologists have continued to improve the databases they use and to employ ever more powerful statistical techniques to analyse the data they contain. This has resulted in several additional estimates of Paleozoic biodiversity, and not just of marine organisms. There have been estimates made of the ancient biodiversity of plants, insects, and mammals. Although I won't be discussing them here, I have included references to them in the Further Reading section at the end of the book. What I do want to review here is a study carried out by John Alroy in 2010.[5] Alroy's study essentially parallels that of Sepkoski. It focuses on the marine invertebrate biota and nicely demonstrates the increasing sophistication of the techniques used in these sorts of studies. It also directly addresses the issue of whether or not the post-Permian expansion of

[5] See Alroy (2010b) in the Further Reading section.

the Modern fauna documented by Sepkoski was real, or a statistical artefact caused by the Pull of the Recent.

Alroy used an online database that allowed easy access to the data (the Palaeobiology Database). Importantly, this database used a different method of recording the data. Sepkoski's original database consisted of a simple tabulation listing the first and last appearances of genera and families with each appearance or extinction event tied to the Geological Time Scale. The database Alroy used is quite different. It lists the genera and families contained within *collections* of fossils as reported in the palaeontological literature. Plus, in addition to the fossil data, each collection has its age, the location it came from, and the sort of rock it was collected from carefully recorded. Because of the unique way that Alroy stored the palaeontological data, he was able to address the non-biological effects that are thought to result in the Pull of the Recent. This is something that Sepkoski was unable to do because his database was unable to make the link between fossils and the rocks they were recovered from.

To minimise the effect of the Pull of the Recent in this analysis, Alroy used a strongly filtered data set. To minimise the non-biological effects, Alroy didn't use collections from poorly or entirely unlithified rocks. Because these rock types are uncommon in older strata, which tend to be harder and more lithified, including these in his analysis would result in an artificial increase in the level of diversity in younger rocks.[6] He also developed a unique approach to how the data was selected from the database that overcame the issues of sampling bias. Alroy was also interested in what was underlying biodiversity, so he filtered his data set in the same way as Sepkoski, removing all the taxonomically uncertain genera and excluding any with species that are alive today. His analysis also identified three evolutionary faunas that were very similar to those documented by Sepkoski.

---

[6] Alroy used the taxonomic level of genus for his study, Sepkoski used the family level.

Figure 4.5 shows Alroy's results, both his estimate of total biodiversity and the three fauna groups he identified: I've labelled them Cambrian, Paleozoic, and Modern. Comparing Alroy's total diversity curve with Sepkoski's estimate (Figure 4.3) is difficult. Firstly, Alroy's analysis is at the taxonomic level of genus, whereas Sepkoski's is at the family level. This may explain the spikier curve in Figure 4.5, as the use of a lower taxonomic level will introduce more noise to the system. Also, because he filtered the data set so strongly, he excluded many genera. As a result, there are only about 350 genera involved at peak biodiversity. Sepkoski, on the other hand, records a peak biodiversity of over 900 *families*, most of which would contain several genera – that's a lot of extra genera.

Despite the 'spikiness' of Alroy's curve, we can still recognise the Big Five mass extinctions. The end-Permian event still stands out as the largest event, but the effects on biodiversity of the other members of the Big Five are greater than those recorded by Sepkoski. This is particularly true of the end-Devonian event, which appears to be close to the end-Permian event in scale. There are, however, additional drops in diversity, for example at the end of the Carboniferous and the end of the Jurassic. These are almost certainly the result of additional mass extinctions over and above the Big Five that were first documented by Raup and Sepkoski.

Alroy certainly succeeded in minimising the Pull of the Recent. There is very little trace of the spectacular rise in biodiversity that is so much a feature of the post-Permian section of Sepkoski's curve. Yes, it's there, but it is minimal. But was he right to filter his data set so strongly to remove the Pull of the Recent? Alroy's methodology and results have serious critics who suggest that in fact he went too far in filtering his data set, which has resulted in masking a *real* increase in diversity. If they are right, the Pull of the Recent doesn't have a significant effect in any biodiversity estimates, and the post-Permian expansion documented by Sepkoski is real.

Where do I stand on this issue? Does the Pull of the Recent play a significant role in the post-Permian expansion or not? I am an

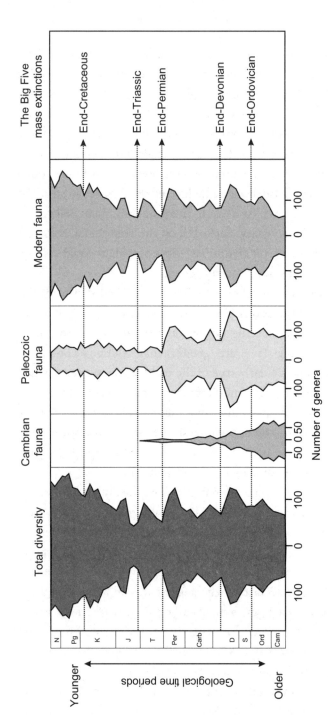

FIGURE 4.5 Alroy's estimates of Paleozoic biodiversity. Axes are the same as previous figures. Note that in this figure the data are at the taxonomic level of genus. The dark grey spindle to the left is the total estimated biodiversity; the remaining three spindles show the changing biodiversity of biotic groups similar to those recognised by Sepkoski. Redrawn with permission from Alroy (2010b).

inveterate fence-sitter,[7] and I believe that this is not an either/or choice. It's hard to argue with statistics, so I find it hard to believe that the Pull of the Recent doesn't play some part in the apparent post-Permian expansion. However, there are several recent studies suggesting that, at least in the case of bivalves (clams, oysters, and the like) and other molluscs, the Pull of the Recent may have relatively little effect and doesn't seriously affect any estimates of diversity. If this result could be shown to be applicable across an entire biota, not just a part of it, then the massive post-Permian increase in Figure 4.2 should be considered, at least in part, to be a real phenomenon – although it should be noted that there is no obvious explanation for *why* this rapid increase in diversity happened.

Looking now at the three evolutionary faunas, the interplay of the diversity and composition of the fauna is obvious. In Alroy's plot, the Cambrian fauna dominates the Cambrian and earliest Ordovician, then fades away, lingering on into younger geological periods than suggested by Sepkoski. The Ordovician expansion recorded by Sepkoski is still there. However, in the Sepkoski plot, this rise in total diversity is caused by a rise in the Paleozoic fauna. Alroy's results suggest that it's the Modern fauna that dominates this rise in biodiversity. In fact, the Modern fauna is the main component of the marine biota from the Ordovician to recent. Regardless of the differences between the two sets of curves, the fact that the same three evolutionary faunas were identified from two data sets using two different methodologies suggests that they are real entities and that they do underlie all ancient biodiversity.

Although all the studies agree on the central role of mass extinctions in the history of life on Earth, a glance at the results of the two studies outlined here shows that there are still significant disagreements. Some of the disagreements might be sorted out if we could directly use species in the analyses instead of genera or families,

---

[7] You will come across several more examples of my fence-sitting as you go through this book.

and if we were able to see the data at a higher temporal resolution. If you look again at Figures 4.3, 4.4, and 4.5, each data point that was calculated sits at the end of each straight line – and some of them are a considerable distance apart, which represents gaps of millions of years. If you think of each data point as a snapshot of diversity through time, these figures represent a stack of snapshots showing changes in diversity. However, instead of a series of snapshots it would be great if we had the video I talked about in Chapter 2, a continuous record of what's going on. A recent publication from a Chinese-led team of palaeontologists has perhaps pointed to a way that will eventually allow us to do just that – visualise these ancient changes in biodiversity as an all-action video.

BRING ON THE SUPERCOMPUTERS

A large group of palaeontologists, led by Jun-Xuan Fan, recognised that in order to achieve the goal of a 'video' of ancient biodiversity rather than a series of snapshots each potentially separated by a considerable amount of time, they would need more data: lots and lots more data. They also recognised that they would need something very powerful to manipulate all that data – so they employed a supercomputer. In fact, they used a fair amount of computing time on one of the world's most powerful supercomputers, the *Tianhe II* at the National Supercomputer Center in Guangzhou, China. The data they used were mined from yet another new database, the Geobiodiversity database.

The Geobiodiversity database doesn't store its data as first and last appearances of taxonomic groups or even as collections, as was the case with the Sepkoski or Alroy databases discussed above. Instead, its data consist of tabulations of marine invertebrate species that were collected layer by layer in a geological section. Figure 2.4 shows a single geological section. The layers of rocks that the fossils were recovered from are plotted down the left-hand side, and every appearance of a fossil is plotted against them. If the section was accurately dated, we could provide an estimate of the age of each fossil appearance. Fan's study used data from 3,000 well dated sections. The sections ranged in age from the Cambrian to the earliest Triassic, a shorter time span than that used by Sepkoski and Alroy. The vast majority of the sections studied were from China, but Fan's data set was supplemented with European fossil data for intervals where Chinese sections didn't yield many fossils.

The use of geological sections rather than presence/absence data or collections used in the palaeobiology database enabled strong links to be made between the fossils and the sort of rocks they were recovered from. In addition, because they were dealing with abundant data from well dated sections, Fan's group could establish the order in which species appear and disappear and give each event a precise age. After carefully vetting the data to make sure the species were all identified correctly, they were left with 11,000 species. They then employed a statistical analysis called CONOP – this technique allows data from across all 3,000 sections to be integrated, stitched together, and the species ranges and the exact sequence of their origination and extinction to be very precisely computed. This approach allowed the ancient changes in biodiversity to be determined at intervals of an astonishing 26 thousand years. That's not a mistake: that's thousands of years. In terms of deep time, that is absolutely amazing.

So what did they find? I have redrawn their data as a spindle diagram in Figure 4.6. It looks a little ragged at the edges, but that's because of the extraordinary resolution being employed. Each tiny wiggle represents 26 thousand years. It certainly gets us closer to the goal of producing a video of species diversity.

The Cambrian Explosion is clear in this plot, and so is the Ordovician expansion, but here it's restricted to the early Ordovician only. The end-Ordovician mass extinction is very clear, but the Chinese data suggest that it is a slower affair than was shown by either Sepkoski or Alroy. The end-Devonian event has become a

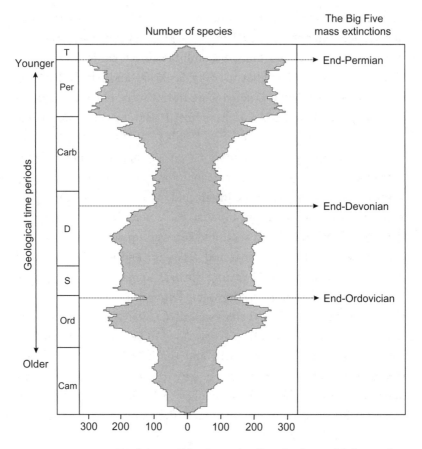

FIGURE 4.6 Cambrian to Triassic species diversity data at 26-thousand-year(!) resolution. Time down the vertical axis, diversity along the horizontal. Redrawn with permission from Fan et al. (2020).

long-term fall in biodiversity followed by an equally long period of relatively low diversity that stretches well into the Carboniferous. In fact, biodiversity really doesn't build up any head of steam until we approach the Permian. The end-Permian event is there, as catastrophic as ever.

This study is undoubtedly a big step forward. It shows what can be done with lots of data and a supercomputer. But there are problems with it: the long period of low diversity that follows the end-Devonian in Fan's plot hasn't been recorded anywhere else. In addition, there are events that are missing, most noticeably a significant mass extinction (not one of the Big Five) that occurs just before the end-Permian event proper. Despite being very well documented elsewhere, it's just not visible in this study. These problems may be due to the relatively low number of species being used – 11,000 may sound a lot, but on a global scale where species are thought to number in the millions, it's only scratching the surface. Also, most of the data come from a single geographical area, albeit a large one. Perhaps that has skewed the results. Nevertheless, I have no doubt that this study provides an approach that offers us a unique opportunity to really come to grips with ancient biodiversity.

I am in awe at what is being attempted in all these studies. Their authors are trying to document the Earth's biodiversity over 500 million years or so of deep time. Think about that for a second. This is a massive task: not only are they peering back into deep time, they are trying to count (at least by proxy) what has happened to the vast majority of species that have ever existed on the planet – remember that this is so large that all organisms alive today are living and dying in the margin of error associated with estimates of total species numbers. While they don't always agree, all the studies discussed in this chapter emphasise the role that mass extinctions play in controlling both the biodiversity of the planet, and the make-up of the biota that existed in each period of Earth history, including our *current* biota.

We need to take a closer look at mass extinctions. There are a number of obvious questions that require answers. At the most basic level, we need to decide what *defines* a mass extinction. Up until now, I've talked about the Big Five mass extinctions and hinted at more, so exactly how many mass extinctions are there? The Big Five are clearly seen in all the biodiversity curves discussed in this chapter. But others are signalled in the data. The mass media are fond of quoting numbers of species going extinct at each mass extinction, but how do we know that information? We would like to know what a recovery from a mass extinction looks like; and above all, we would dearly love to know what causes them. This is especially relevant to humans – and everything else living on the planet – right now.

5   Mass Extinctions: The Basics

A few kilometres northeast of the Italian medieval town of Gubbio, just as the road enters the Bottaccione Gorge, there's an excellent restaurant – the Osteria del Bottaccione. Situated in a small, isolated, ancient-looking building, it serves delicious traditional food washed down with strong local wine. You can round off your meal with a delightful, homemade, herbal, and definitely alcoholic, digestif. If you ever visit the osteria (and I do recommend it), have a look at its amazing visitors' book. In it you'll find signatures of famous geologists and palaeontologists from around the world, records of visiting scientific groups, along with oddly drawn pictures of dinosaurs with meteorites looming over them. The reason for all this scientific interest (apart from scientists' great love of good food and drink) can be found a little further into the gorge, at a site that has been helpfully signposted by the Gubbio tourism authority. Here the very youngest Cretaceous rocks are directly overlain by the oldest Paleogene rocks, offering a complete record of the end-Cretaceous mass extinction.

Figure 5.1 shows a close-up of the part of the section that contains the boundary. The deep cleft running more or less diagonally across the image marks the boundary between the Cretaceous (the sheets of rock below the cleft) and the Paleogene rocks (the slightly more massive material above the cleft). To put it another way, what you are looking at is a record of the diversity of life in the Cretaceous, leading up to the end-Cretaceous mass extinction, when up to 75% of all marine species on Earth died. The mass extinction is then followed by the evolutionary refilling of the emptied ecological niches in the earliest part of the Tertiary.

FIGURE 5.1 The Gubbio section – a complete record of the end-Cretaceous
mass extinction (photograph by the author).

But the importance of the Gubbio section lies not just in its
completeness (nor in its proximity to a great restaurant), but in what
was discovered in the rocks within the cleft. The rocks on either side
of the cleft are limestones: hard calcium carbonate made from the
fossilised remains of marine animals with shells. Sandwiched
between the two – at the exact position of the mass extinction – is a
layer of soft clay, free of calcium carbonate. Because it's softer than
the surrounding limestones, the clay layer preferentially weathers
away, leaving the cleft. In recent decades, the weathering process
has been assisted by many scientists poking their fingers in and trying
to collect some of the clay under their fingernails as a souvenir.

This pattern of limestone/clay/limestone suggests that at the
precise moment of extinction, the carbonate supply was turned off,
and it was not turned back on again until sometime within the early
Paleogene. This is perhaps not surprising – the majority of the end-
Cretaceous extinctions were of marine animals (despite dominating

the public's mind, the dinosaurs are only a relatively small part of the extinction event). If we are trying to understand the cause of the extinctions – and understand how diversity recovered afterwards – then knowing how long the carbonate supply was turned off is critical. The complete section at Gubbio offers the opportunity to address the question.

A geologist, Walter Alvarez – together with his father, Nobel prize-winning physicist Luis Alvarez – decided to tackle the problem. They came up with a brilliant idea for dating the carbonate gap involving the element iridium. They assumed that there was no source of iridium on Earth,[1] which means that any that we find here has rained down on the planet from outer space. Not only that, but the rate at which it arrives on Earth (its flux) has been accurately measured. This steady rain of iridium accumulates in depositional basins such as oceans where it is preserved in sediments like the limestones and clay of the Gubbio section. While calcium carbonate is being produced and deposited as limestones, the iridium flux should be diluted in the accumulating sediment, so low levels of the element should be recorded in the limestones. But when the limestone production is turned off – as appears to have happened while the clay layer was being deposited – the flux is no longer being diluted, so iridium levels in the clay should rise. This suggested to the Alvarez team that if they measured the amount of iridium in both the limestones and the clay, then, knowing its flux from space, they could work out how long the carbonate-free period lasted.

In addition to the section at Gubbio, they tested their idea on two other places in widely separate parts of the world: Woodside Creek in Aotearoa/New Zealand and Stevns Klint in Denmark, both of which also contain a clear boundary clay layer at the end of the Cretaceous. As was predicted, in all three sections, samples from the

---

[1] Not exactly true – the Earth's mantle also contains iridium, and that can be transferred to the surface by volcanoes. However, there are no volcanic deposits in the Gubbio section.

clay yielded higher iridium levels than the surrounding limestones. But there was a problem: there was too much iridium in the clay from all three places. Far, far, too much – certainly more than could be explained using the measured background flux alone. To the Alvarez team, this suggested that extra iridium had been added to the planetary system at precisely the same time as the end-Cretaceous mass extinction. And they didn't think that the synchronicity between the addition of iridium and extinction could be a coincidence. In 1980, they published a model that blamed the end-Cretaceous mass extinction on the impact of a huge meteorite (about 10 kilometres in diameter), which delivered iridium – and death – to the planet. The model suggested that the clay layer at Gubbio (and the other sections) was largely composed of the microscopic debris of the meteorite (no wonder eager Earth scientists try to scrape up a fingernail full of the stuff as a souvenir).

I was a young PhD student at the time, working on late Cretaceous microfossils, and I remember being absolutely shocked at the audacity of this model. At that time, the thought of a geological event caused by an extra-terrestrial event was almost unthinkable: geologists just hadn't been trained to believe that extra-terrestrial forces could influence the Earth's geological history. It's true that some extra-terrestrial causes had been discussed before – cosmic rays had been suggested as a killing mechanism for the dinosaur extinction – but they amounted to little more than untestable speculations. The meteorite impact model was based on hard evidence and was eminently testable. I admit that to begin with I struggled with the idea, and it took the later arrival of more detailed evidence to convince me that the impact had really occurred. But that evidence came quickly – geological, geochemical, and geographical. The identification of a 200-kilometre-diameter crater under Chicxulub on the Yucatan peninsula, Mexico that was dated to be of end-Cretaceous age – and not of volcanic origin – was certainly compelling. It took longer for others to accept the model: when I first joined my university, I tried to put up a display on the end-Cretaceous meteorite event, and one of my colleagues objected. He thought the whole concept wasn't proven science but pure speculation. It is now mainstream science, with pretty much all scientists accepting the strong evidence for the impact. However, as you will see below when I discuss possible causes of mass extinctions, although I fully accept the evidence for a meteorite impact, I'm not convinced that it explains every aspect of the end-Cretaceous mass extinction event.

The excitement following the release of the Alvarez impact model resulted in a detailed re-examination of all mass extinctions. In particular, the hunt was on for evidence of extra-terrestrial causes for all of the Big Five mass extinctions. A veritable flood of papers examining all aspects of the events poured into the literature. The picture of mass extinctions that emerged from this re-evaluation turned out to be much more complex than first thought, and things are far from settled even now: we can't even agree on how many mass extinction events occurred in Earth history, let alone what caused them all![2]

WHAT EXACTLY IS A MASS EXTINCTION?

Before getting into details of possible causes, we need to cover some basics. Firstly, I need to define what a mass extinction is. Secondly, I need to come to some sort of decision about how many there were.

---

[2] Before we move on, a brief note about terminology and my own personal prejudice. Throughout this book, wherever possible I refer to each mass extinction as 'end' followed by the geological period associated with it: end-Permian, end-Triassic, etc. This isn't the usual way to name mass extinctions, although it's by no means unheard of. The normal way is to name them after the geological boundary they are associated with: the Permian/Triassic (often shortened to P/T) extinction event or the Triassic/Jurassic (Tri/Jur) extinction event, for example. The end-Cretaceous extinction event was known as the Cretaceous/Tertiary event, shortened to K/T (K being the conventional geological shorthand for the Cretaceous, as C is already taken by the older Carboniferous Period). The 'K/T' event had a good ring to it and was well accepted. Unfortunately, the geological powers-that-be decided to do away with the Tertiary period altogether, and replace it with two others, the older Paleogene followed by the younger Neogene. Thus, the K/T mass extinction event became the K/Pg event, which has less of a ring to it (in fact it sounds to me like a group of accountants). So I prefer not to use it, and instead I use the end-Cretaceous – and to be consistent, wherever possible I'll use the same style of terminology for all mass extinctions.

The definition of mass extinction I am going to use is a simplified version of one suggested by Sepkoski in 1986:

> A mass extinction is an increase in the level of extinction over a very short interval of geological time of more than one geographically widespread higher taxon, resulting in a drop in biological diversity.

This definition is deliberately somewhat vague, but there are rules to use when it's applied:

(1) The level of increase in extinction must be above the background level of extinction at the time of the possible mass extinction, but the level of increase is not specified.
(2) There is no indication of the length of time over which the extinctions should occur. But, within the constraints of the geological time scale, it must be relatively rapid.
(3) The effect of the extinction event has to be global and not just local to one particular region. This means that the taxa going extinct have to be geographically widespread. The number of taxa going extinct for it to be a *mass* extinction is left open, but clearly it has to be many more than one.
(4) The taxonomic level we use to decide if an event is a mass extinction is also left unspecified. Although much of the documentation of mass extinctions is based at the level of genera, there is no doubt that in all cases the extinctions involve the loss of even higher-level taxa such as families or orders. Whatever taxonomic level is picked, when dealing with the fossil record it won't be at the species level; species are notoriously difficult to count accurately, for reasons outlined in Chapter 2.

Why the vagueness in the definition? Why not just say that a mass extinction is defined, for example, as a 'fourfold increase in extinction of three families occurring over 0.5 million years'? That would be a workable definition. But what about events with only a threefold increase, that occurred over 0.6 million years? They would be global and big, but, using the tight definition, not quite a mass extinction. Best to be a little vague to try to catch all of the major mass extinction events in Earth history.

A precise definition could also exclude some possible triggers for mass extinction. For example, in 1984 Andy Knoll suggested that a mass extinction should be 'instantaneous when viewed at the level of resolution provided by the geologic record'. This definition would only include extinction events that have no precursors in the fossil record – those with an 'alive today – dead tomorrow' scenario, such as meteorite impacts. It would exclude any event that was longer in duration, such as those with a longer-term global environmental trigger.

So – armed with our somewhat vague definition – we can look at how many mass extinction events have occurred in the Earth's history. Raup and Sepkoski's original analysis (Figure 2.3) suggested five. These are the so-called Big Five mass extinctions popularised in the media. I use the term 'Big Five' with some personal scepticism. This is because more recent analyses carried out using the new data sets that I talked about in the global diversity chapters have muddied the waters somewhat. Figure 5.2 shows the results of one such study, by John Alroy. To produce this curve, Alroy used the sophisticated data set we introduced in Chapter 4. Unlike the Raup and Sepkoski extinction curve (Figure 2.3), which presented the data at the taxonomic level of the family, Alroy's analysis is at the genus level. The rate of extinction is the number of genera disappearing per interval of geological time.

To start to understand the diagram, first try to ignore the peaks and troughs and, in your mind, draw a line of best fit through the data from left to right. There is a general trend there – a line angled slightly downwards from left to right – and it's the same one we saw in the original Raup and Sepkoski curve (Figure 2.3) where 95% of the extinctions were confined to a region that also had a slope downwards from left to right. This means that this new analysis again suggests that, over the last 500 million years, the rate at which genera are going extinct appears to be decreasing. Overall, it's getting easier to survive!

Now, look at the details of the diagram. The heights of the peaks are very varied. The fact that there is a spectrum in the size of

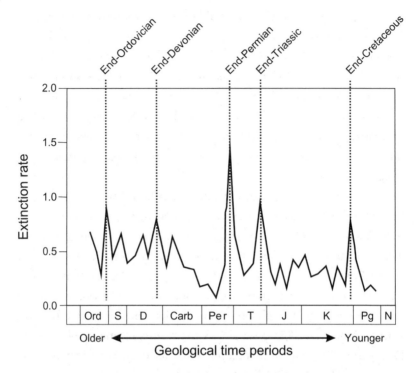

FIGURE 5.2 Rates of extinction over the past 500 million years. The horizontal axis is geological time. The vertical axis is the extinction rate of genera per geological time interval. The symbols represent the various geological periods. The Big Five mass extinctions are indicated by the dashed lines. Adapted with permission from Alroy (2008).

extinction events has led some to suggest that mass extinctions are not real, but simply the result of variations in the background extinction rate. But this is misleading. The very definition of a mass extinction event requires the extinction rate to stand out above the background rate either side of the event: in other words, they must be seen in the context they are set in.

Figure 5.2 also makes it clear that there have been far more mass extinction events than just five. In fact, the curve is quite jagged. We saw the same thing in Alroy's diversity curve (Figure 4.5), where I hinted that there were more than five mass extinctions. In

Figure 5.2, I can count about thirteen or fourteen peaks in extinction rate that would probably satisfy our definition of mass extinction. And it gets worse: this is a large-time-scale diagram. If we could zoom in on parts of it and look more closely at certain time periods, we would see even more extinction peaks. For example, in detail, the end-Permian mass extinction event probably involves a number of mass extinction events that are very close to each other in time (geologically speaking, that is: in fact, they are spread out over about 10 million years). The oldest of these extinctions occurs during the Capitanian age (see Appendix 1 for the Geological Time Scale) and is almost as significant as the end-Permian event itself.[3] However, at the scale of Figure 5.2 we lose this detail and complexity, and all we see is a single peak. Other geologists have suggested that there are even more candidates that could be considered as mass extinctions.

This observation is one of the reasons for my dislike of the term Big Five, since clearly there have been a far greater number of mass extinctions than five. But what about extinction intensity? Are the Big Five special in that respect? Are they the *biggest* of all the extinction events? Well, no. In Alroy's diagram of the Big Five extinctions, only the end-Permian, end-Triassic, and end-Cretaceous events stand out as being really exceptional. The end-Ordovician and end-Devonian extinctions – although the biggest of the pre-Permian extinctions – seem to get a bit lost amongst the noise of the surrounding extinction peaks. There are extinction events that are also great candidates to join the 'Big' group, but because they were recognised after the Big Five were labelled, they missed out. The Capitanian extinction is an obvious example.

My opinion of the Big Five? I'm convinced that the extinctions at the end of the Permian, Triassic, and Cretaceous are singular enough to be included in an *exceptional* group. I am, however, much less convinced by the extinctions at the end of the Ordovician and

---

[3] This is the extinction event that seems to be missed in the study by Fan and his team in Figure 4.6.

Devonian: they are the largest of the Paleozoic extinction events, but probably not by much if Alroy's data are to be believed. Nevertheless, despite this and the presence of many additional events, I'm going to continue to use the term 'Big Five'. They were the original set of mass extinctions recognised by Raup and Sepkoski, and the term is in common usage, so it provides a convenient shorthand for me. So, for better or worse, I'm going to concentrate on these five events in the rest of the book. But we should remember that this term is a vast oversimplification, hiding the complexity of the real number of mass extinctions, and their effects on global diversity.

HOW BIG ARE THE 'BIG FIVE', AND HOW DO WE KNOW THAT?

Up until now, we have been dealing with the number of genera or families going extinct. This makes it a little difficult to appreciate the scale of these events. It would be much better if we could work out the number of species that went extinct (because genera or families contain wildly different numbers of species). This is extremely difficult to measure directly. We cannot simply count the number of species dying out: the fossil record doesn't record that amount of detail. So, an approach had to be developed that allows species-level estimates to be made based on what we do have good data for – the number of genera or families involved.

It's hard to kill off a whole genus, let alone a family. There are four genera and seven species belonging to the family Hominidae (see Table 2.1). They are the genus *Homo* (one species: us) and genera containing chimpanzees and bonobos (two species), gorillas (two species), and orang-utans (two species[4]). So, if we are to remove the family Hominidae from the Earth, to cause this family to go extinct, we must kill off every living individual from all seven species. That may seem not particularly difficult for Hominidae; sadly, we're well

---

[4] There has been a suggestion of a third orang-utan species (see Nater and others (2017) in the Further Reading section) – but this remains controversial.

on the way with gorillas and orang-utans – although getting rid of 7–8 billion individual *Homo sapiens* may be a bit of a challenge.

A thought experiment suggested by David Raup emphasises exactly how difficult it is to cause a higher-level taxon to go extinct. He called his experiment 'The Field of Bullets'. Imagine a world with a single phylum. It has ten classes, each with ten orders, each with ten families etc., down to ten species, each of which contains only ten individuals: this still results in a million individuals. Put all these individuals into a field and randomly shoot 75% of them – the randomness is important. After getting rid of 75% of individuals, what effect (statistically) does that have on the higher taxa? The likelihood of losing the phylum is 0% – as long as there is one individual alive, the phylum still exists. Each class has 100,000 individuals, so the chance of a whole class going extinct is also practically zero. And so on down the list we can go. Raup estimated that the chance of knocking out a whole species is only about 5%. That means if we could kill off 75% of all individual animals and plants from anywhere on the face of our planet, absolutely at random (and I admit that is a horrifying thought), then we have only a 5% chance of killing off a species, which means that the chance of making a genus or a family extinct is less than 5%. Thankfully, this is only a thought experiment, but it did suggest that we could model the number of a higher-level taxa that would go extinct, if we knew how many species had been killed off. In the case of a mass extinction, however, we want to run the model in reverse. We need to estimate species-loss based on the number of genera of families that go extinct. In order to do this, Raup used what are known as 'rarefaction curves'.

I have drawn some rarefaction curves in Figure 5.3. Curves like this are used by modern ecologists to estimate the probable number of species, genera, and other higher-level taxa present today from a count of individuals in an area. When used this way, the horizontal axis records the number of individuals counted and the vertical axis the possible number of higher taxa present (Figure 5.3a). Say, like Darwin, we are interested in beetles, and in our study area we counted sixty

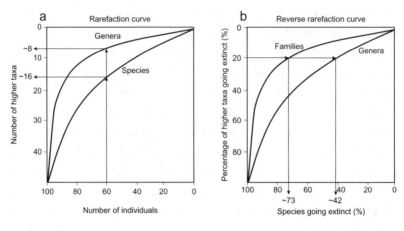

FIGURE 5.3 A pair of imaginary rarefaction curves used to estimate (a) the number of higher taxa present in a study area and (b) the number of species going extinct.

individuals. From the sixty on the horizontal axis, we draw a vertical line until it intersects first the lower curve. This curve allows us to estimate the probable number of species present. We do this by drawing a horizontal line from the intersection point across to the vertical axis and reading off the number. In the example here, a count of sixty individuals suggests the presence of about sixteen species. Similarly, if we want to estimate the number of genera present, we continue our vertical line upwards until we intersect the upper genus curve, draw our horizontal line across to the vertical axis and read off the result. In this case, the count of sixty individuals suggests that they represent about eight genera.

There are two factors that are important to get right if this is going to produce realistic estimates and allow us to apply the Field of Bullets model. Firstly, we have to collect our beetles randomly. If we don't do this, we will significantly reduce our confidence in the final estimate. For example, if we were interested in the overall diversity of beetles in a forest and we carried out our count from a single location, how can we relate that to the beetle diversity of the whole forest?

What would happen if we collected our sixty individuals from a single dead tree? With sampling limited to a single ecological niche, we would expect to see only a few species, but the curve will tell us that there are sixteen. To produce good estimates of higher taxa, what is needed is a careful, random sampling programme that encompasses the entire forest.

The second important factor is the shape of the curves themselves. These are documented by patient biologists who will go out into the field and, over time, document the relationship between the number of individuals, species, and genera. In the beetle example, this curve would probably be the result of many studies carried out in various forests. Each rarefaction curve will apply to one animal group – we shouldn't apply curves derived from an analysis of beetles to work out the number of species or genera of butterflies, for example. A separate set of curves needs to be calculated for each animal group we are dealing with.

To obtain an estimate of species extinction for each mass extinction, Raup suggested that we could run the rarefaction curve in reverse. The Field of Bullets scenario suggests that if the killing that occurs during a mass extinction is truly random, there is a statistical link between extinction at species and higher taxonomic levels. But the extinctions have to be random. This is analogous to the random sampling needed to ensure that our estimates of beetle diversity were realistic. The reverse rarefaction approach is shown in Figure 5.3b. Note that the axes have changed: we now have the percentage of species going extinct on the horizontal axis, and the percentage of higher taxa going extinct on the vertical axis. The curves have slipped up a taxonomic level.

In the case of fossils, deciding on the shape of the curves is an obvious problem. We should be drawing a different set for each animal group going extinct. However, when it comes to fossils, there are very few groups where we know with any certainty the number of species in the various genera and families. So Raup had to choose a group of animal fossils where the species/higher taxon relationship was well understood and use that as a *proxy* for all other animal groups. The group he chose to use was the echinoderms, a marine group that includes sea urchins, sand dollars, starfish, and sea cucumbers. It's important to note that because of his choice of animal to define the curve, the estimates that Raup came up with are for marine species only. Even though we have a very good record of the echinoderms, it would be stretching things a bit far to apply the curve to terrestrial animals and plants.

The process of using Figure 5.3b to obtain species-level extinction estimates is just the reverse of the rarefaction approach I explained above. As an example, if 20% of families go extinct at a mass extinction event, we draw a horizontal line from the 20% extinction point on the vertical axis until it intersects the family curve. From that intersection point, we drop a vertical line down to the horizontal axis and read off the number of species going extinct: in this case we would be looking at a species extinction rate of around 73%. If we lost 20% of all genera, we would start at the same extinction point but extend the horizontal line until we intersect the genus curve, then drop a vertical line down to the horizontal axis, and this suggests a lower level of species loss – around 42%.

Based on this approach, estimates of species-level extinctions for each of the Big Five are shown in Table 5.1. Remember that these are marine extinctions only[5] and that they are based on the Sepkoski data set. Nevertheless, there are some frightening numbers in the table – it suggests that at the end-Permian mass extinction event about 95% of all marine species on Earth went extinct, and all the data we have available suggest that they disappeared over a very short length of time.[6] The remaining Big Five are also substantial: 84% species loss at the end of the Ordovician; and 70–78% at the end of the Cretaceous. But how good are these estimates?

---

[5] Some, of course, might argue that they are really only applicable to echinoderms.

[6] Estimates for the duration of the end-Permian event range from 20 thousand to about 50 thousand years.

Table 5.1 *Estimates of the level of marine species extinctions at each of the Big Five based on the percentage of family or genera going extinct*

Data from Jablonski (1994).

Mass extinction	Estimated species loss based on family-level extinctions	Estimated species loss based on genus-level extinctions
end-Ordovician	84 ± 7%	85 ± 3%
end-Devonian	79 ± 9%	83 ± 4%
end-Permian	95 ± 5%	95 ± 5%
end-Triassic	79 ± 9%	80 ± 4%
end-Cretaceous	70 ± 13%	76 ± 5%

There are two obvious problems with the reverse rarefaction approach:

(1) *It's all based on echinoderms.*

The curves in Figure 5.3 were constructed using data based on our understanding of the taxonomic structure of the echinoderms, which was then applied to all other animal groups. This may not be appropriate – it is a near certainty that other animal groups won't have the same distribution of species/genera/families as Echinoderms. There is no easy way of checking this. However, Raup did point out how well the estimates of species loss from families and genera agree. He argued that if the distribution curves were way off this probably wouldn't be the case.

(2) *The killing has to be random.*

The reverse rarefaction method relies on the Field of Bullets scenario, and for that to work, the killing has to be random. Unfortunately, we know that the killing that goes on in mass extinctions is anything but random. Instead, what happens is that although significant proportions of ecosystems are devastated, the effects are commonly concentrated in certain areas. As we will see shortly, the end-Permian mass extinction was particularly severe on reefs, effectively removing them from the planet. Because we are removing a whole reef ecosystem, the killing will not be random. It's like the example of collecting beetles from a single tree when we are interested in the diversity of a whole forest. This non-random extinction will probably lead to an overestimate of species-level extinctions.

So where does that leave us? The data in Table 5.1 still represent the best estimates that we have for the level of marine species-level extinction in each of the Big Five. But we should take them with a significant pinch of salt. At the worst, the whole basis for the reverse rarefaction curves is wrong, and our estimates are way off (possible but unlikely). At the best, they are overestimates because the killing simply wasn't random.

CYCLES OF EXTINCTION

Ever since they were first clearly identified by Raup and Sepkoski, there have been heated discussions about whether there is a pattern to the occurrences of mass extinction. Do they occur at regular intervals? Is there a rhythm to the death and destruction that they cause? There are, after all, many cycles in nature. Every day, cycles beat through all of the natural world, affecting every living thing. The regular ebb and flow of the tide and the passage of the seasons are obvious cyclical changes to the environment. Such natural rhythms influence life on Earth on a daily, yearly, or decadal time scale. And there are longer-term beats – ones that operate over deep time. For example, in the 1920s, the Serbian scientist Milutin Milanković documented three astronomical cycles: one associated with passage of the Earth around the Sun (a 100,000-year cycle), another with the degree of the tilt of the Earth's axis (a 40,000-year cycle), and finally a cycle based on what is known as the precession or the rotation of the axis relative to fixed stars (about a 20,000-year cycle). Beating together, these three cycles influence the amount of sunlight falling on the planet. Although the variation in sunlight is small, because the Earth has a tilted axis, its effect is strongest at the poles. Over the aeons of deep time, this small variation in sunlight at the poles makes enough of a difference to drive the advance and retreat of glaciers during the ice ages. It's hard to overestimate the changes that this rhythmic pulsing of ice sheets across continents would have made to the biota living at the time. These subtle Milanković cycles are also strongly implicated in the development, about 34 million years ago, of the massive ice sheets that characterise Antarctica today.

Could long-term regular cycles like the ones documented by Milanković be reflected in changes in the diversity of life on Earth? Palaeontologists became interested in the possibility that mass extinctions were cyclical, visiting death on the planet with clock-work regularity. Although speculation about possible cyclicity increased soon after Raup and Sepkoski documented the presence of the Big Five, studies of possible natural cycles of diversity go back a lot further.

In the 1940s and 1950s, using the limited data available and an old Geological Time Scale, palaeontologists recognised repeated peaks in diversity resulting from what were called 'bursts of evolution' or 'accelerated evolution'. In 1952, Norman Newell was able to docu-ment cycles of accelerated evolution preceded by intervals of low diversity in a wide variety of fossil groups. Today we would link the low-diversity intervals with mass extinctions, but he referred to them as the periods of 'extensive extinction' (although later he would be one of the first palaeontologists to use the term *mass extinction*). Newell suggested that the periods of accelerated evolution that followed each interval of low diversity were a response to the many new ecological niches made available by the episodes of extensive extinctions that preceded them. At the time, the linkage between the units of the Geological Time Scale and their numerical ages was imprecise, so he could not be certain about the exact timing for his alternating intervals of high and low diversity. Nevertheless, he attributed them to some form of recurring natural geological cycle of uplift and mountain building. This was never going to be a full explan-ation of these alternating intervals of high and low diversity. It just changed the question from 'What caused cyclic changes in diversity?' to 'What caused the regular cycles of mountain building?'

Since Newell's work, our understanding of the fossil record has steadily improved, and a study published in 1977 by Alfred Fischer and Michael Arthur demonstrates how far it had progressed by that stage. They documented cycles in diversity much more precisely than Newell. Fischer and Arthur focused their study on the abundance of marine fossils from the latest Permian to the present day. They recognised seven intervals of high animal diversity that appeared to coincide with warmer oceans and high sea levels. They called these intervals 'polytaxic'. Separating the polytaxic intervals were seven shorter periods, marked by low faunal diversity and colder water. These intervals were also notable for sudden blooms of individual organisms, members of what are often referred to as disaster species that appeared to thrive in conditions that other species could not tolerate. They called these intervals 'oligotaxic' and suggested that episodes of what they termed 'biotic crises' (again we would call them mass extinctions) occurred during these oligotaxic intervals. Within the limits of the Geological Time Scale being used, they found a strong rhythm to the changes in oceanic conditions that they had documented. The switch from oligotaxic to polytaxic periods appeared to occur every 32 million years. Unlike Newell, Fischer and Arthur rejected any role for geological cycles of mountain building. Instead, they suggested that rhythmic changes in oceanic circulation, driven by changing climates, were responsible. But again, this just shifts where the explanation is needed from 'What caused the cycles of diversity?' to 'What caused changes in the ocean circulation every 32 million years?'

In 1984, Raup and Sepkoski applied their own considerable statistical skills to the search for cyclical patterns in the fossil record. They had some advantages over Fischer and Arthur: notably the extensive Sepkoski data set (which had already proved its worth in the identification of the Big Five mass extinctions) and a more precise Geological Time Scale. Unlike Fischer and Arthur, Raup and Sepkoski didn't address the issue of overall diversity changes; instead, they focused entirely on the timing of mass extinctions. They did, however, use the same study interval, the period between the end-Permian event and the present day. Based on their statistical analysis of the Sepkoski data set, they suggested that over this time interval a mass extinction (beyond the Big Five) occurred every 26 million years (later refined to 26.2 Myr). The precision of their result and its statistical backing rocked the palaeontological community, creating intellectual waves that ripple through to the present day. Think about it: what could cause a regular periodicity to mass extinction events?

Because of uncertainties surrounding the exact date and the size of each mass extinction event, some palaeontologists immediately dismissed the perceived cyclicity as being merely a statistical artefact, a signal produced by Raup and Sepkoski's calculations that was not in the original data and therefore not real. Undeterred by such criticism, Raup and Sepkoski remained very confident that they had identified a clear rhythm to life; or in the case of their study, a rhythm of death.

> It seems inescapable that the post-Late Permian extinction record contains a 26 Ma periodicity...
>
> *(Raup and Sepkoski, 1984)*

A mass extinction every 26 Myr? What could drive this sort of clockwork extinction? Raup and Sepkoski couldn't see any geological, biological, or climatically induced events that could drive a cyclicity of 26 Myr, so they looked to the heavens and sought an extraterrestrial cause. Soon, physicists, astrophysicists, mathematicians, and statisticians joined the discussion. Most of these scientists – together with Raup and Sepkoski themselves – favoured a model that linked the 26-Myr extinction cycles with the movement of the Sun in our Galaxy.

As the Sun circles our disk-shaped Galaxy, the Milky Way, it moves up and down through the main body of the disk, the Galactic Plane. The suggestion was that as the Sun crosses the Galactic Plane, 'something happens' that results in a mass extinction on Earth. The

Sun's crossing takes place every 32–42 Myr. While this timetable isn't even close to the suggested 26-Myr cycle of extinction, many (but not all) workers thought that the uncertainties associated with the dating of the extinction events used in Raup and Sepkoski's calculations were so large that the 26 Myr could be stretched to fit this longer cycle.

But what was the 'something' that happened as the Sun passed through the Galaxy's disk that triggered a mass extinction on Earth? Some speculated that the passage through the disk increased the flux of cosmic rays reaching the surface of the Earth to the level where they caused mass extinctions. This wasn't a new idea. In 1950, the German palaeontologist Otto Schindewolf had suggested that a burst of cosmic rays from a nearby supernova caused the dinosaur extinctions, for example. However, if an organism is covered by a large amount of water, it will be protected from damage caused by the impact of cosmic rays. As we have seen, a significant proportion of the extinctions associated with any mass extinction occur in the marine realm, and there the organisms would be heavily shielded from damaging rays. So, although a burst of cosmic rays could possibly explain terrestrial extinctions, it couldn't account for the marine extinctions.

Another model suggested that galactic dust was concentrated in the Galactic Plane and, as the Sun and Solar System moved through it, the dust could severely dilute the amount of sunlight reaching the Earth's surface. Indeed, some suggested that the dust may have blocked out the Sun altogether, plunging the Earth into global winter. This is a possible kill mechanism, and the impact model for the end-Cretaceous extinctions – which we will discuss in more detail in the next chapter – suggests that a drastic change in the level of sunlight is a possible trigger for ecosystem collapse and could cause a mass extinction. Ultimately, however, the lack of any observable increase in the density of dust along the Galactic Plane has led to this model being rejected as well.

Perhaps it was all to do with extra-terrestrial impacts. Raup and Sepkoski's work on cycles of extinctions was published soon after the

Alvarez impact model for the end-Cretaceous event. Even though only the end-Cretaceous event was clearly linked with an impact – the iridium layer and identification of a crater created at the right time are good evidence – it seemed possible, or even likely, that eventually evidence would be found to link all mass extinctions with the impact of an extra-terrestrial object.

Later, statistical analysis of the age of impact craters provided circumstantial support for the idea. The analysis used only the largest craters (those thought likely to have been caused by the impact of an extra-terrestrial body big enough to trigger an extinction event), and these appeared to be clustered in time around each of the Big Five mass extinction events. However, there are problems when carrying out a detailed analysis of meteorite craters over deep time. Firstly, we can't be certain that we have found every crater that might trigger a mass extinction – even though we are only considering large ones. Weathering can erase evidence for a crater; they can be buried under kilometres of younger sediments, or removed through plate tectonic activity (subducted away). Secondly, it is very difficult to put a precise age on a crater. It can be estimated if we know the ages of both the rocks that were hit by the meteorite, and the rocks that fill the crater, but it is an imprecise art. The further we go back in time, the harder it is to accurately identify and date large impact craters.

And even if we could prove that every mass extinction was a result of an impact (which they don't seem to be), we would be faced with the same problem as we found in the early pioneering work of Newell and of Fischer and Arthur. It simply shifts the place where an explanation is needed. If all mass extinctions were caused by impacts, what caused a meteorite or other extra-terrestrial body to impact the Earth, like clockwork, every 26 million years? Again, the Sun's movement through the Galactic Disk was thought to be an obvious culprit. It was suggested that the gravitational effects of the Sun's passage through the Galactic Plane triggered a flood of meteors or comets into the Solar System. Some of these impacted on the Earth, resulting in a mass extinction. That explanation seemed to cover everything: if all the mass extinctions were a result of impacts, then this model provided the clockwork.

But the mismatch between the 26-Myr extinction cyclicity and the 32–42-Myr time period between each passage of the Sun through the Galactic Disk really started to worry many scientists. In the end, the inability to get these dates to match in any sort of convincing way led to explanations based on our Sun's movement through the Galactic Disk falling out of favour. Scientists again started looking for a new, extra-terrestrially based event that (a) triggered an impact causing a mass extinction and cratering, and (b) had a cycle of about 26 Myr, with no stretching of ages to fit.

Enter *Nemesis*.

THE RISE AND FALL OF NEMESIS

During the 1980s, an object that came to be known as Nemesis[7] was proposed as an 'unseen companion' that orbited the Sun in an extremely elliptical fashion. Every 26 million years, Nemesis looped close to the Solar System, passing through the Oort cloud, a volume of space surrounding the Solar System that is densely packed with icy planetesimals. The gravitational effects caused by the passage of Nemesis disturbed these objects, sending millions hurtling into the Solar System as comets. Some of these hit the Earth, triggering a mass extinction. In one fell swoop, Nemesis provided a mechanism for triggering impacts and a perfect match for the cyclicity – in fact, the orbit could be adjusted to fit any cyclicity. Never mind that we had never seen such an object or found any evidence for it, such as gravitational anomalies. For many, Nemesis was the answer.

Calculations based on the gravitational effects needed to trigger a shower of comets suggested that Nemesis was a type of star called a brown dwarf, the cooled remnants of a very bright white dwarf. Because it is cool, a brown dwarf doesn't give off light, and hence

---

[7] It is named after the Greek goddess Nemesis. She is the goddess who extracts retribution for arrogance before the gods.

it's hard to detect (in the case of Nemesis, impossible to detect!). There were scientific doubts about Nemesis, and we'll talk about them shortly, but they weren't shared by the general public. Nemesis caught people's imagination from the outset. There were articles in the popular press prophesying the return of a 'death star' bringing with it doom and destruction. Today, Nemesis is hopelessly intertwined with a mythos surrounding another extra-terrestrial harbinger of death and destruction: Nibiru.

If you are at a loose end on a very rainy day and feel like wasting some time (and you will be wasting time), go to your favourite internet search engine and type in 'Nibiru'. Click on any link that gets thrown up, and you will enter an alternative reality involving planetary destruction, the new world order, the end of days, the Book of Revelation, and various species of aliens. Linking all these bizarre notions is a planet (or dwarf star or binary system) called Nibiru. Sometimes referred to as Planet X, this object moves around the Sun in a highly eccentric orbit occasionally coming close to the Earth, leaving a wake of destruction. If you dare, keep on clicking and see how deep the rabbit hole goes.[8] It's easy to see why the Nemesis model was so eagerly accepted by the devotees of Nibiru. It appeared to offer real scientific support for this life-ending extra-terrestrial visitor. I honestly don't know (neither do I have the will to chase through all the Nibiru sites) what came first, Nemesis or Nibiru, but today on some websites the two bodies are synonymous.[9] Ironically, as the conflation of Nemesis and Nibiru gained momentum in the general public, the whole idea of Nemesis was fading scientifically.

---

[8] The most unbelievable thing I saw on one of these sites was a 'photo' showing Nibiru looming large above New York city, looking a lot like something out of Lars von Trier's movie *Melancholia*. The text that accompanied the image breathlessly claimed that Nibiru was here, doom was upon us – it was just that no one in New York had noticed it yet.

[9] I sometimes wonder what would have happened if the scientific community hadn't named the thing Nemesis. Would it have caught the public imagination with a less exciting name?

From a palaeontological perspective, the fatal blow for the Nemesis model is that only the end-Cretaceous extinctions can be tied explicitly to an impact event. Any model attempting to explain cycles of mass extinctions based on repeated extra-terrestrial events demands that every extinction event is a result of the same process. Whether it be impacts, cosmic rays, or the Earth's passage through cosmic dust clouds, all mass extinctions have to have the same extra-terrestrial cause. Modern studies have clearly demonstrated that this is simply not the case. The only extinction event linked with an extra-terrestrial event is the one at the end of the Cretaceous.

Many astrophysicists also expressed doubts about the Nemesis model. The proposed highly elliptical orbit was thought by some to be very unstable and would eventually result in Nemesis being unable to maintain the precise 26-million-year cycle needed to match the sup-posed repeat timing of the mass extinctions. It was even possible that the orbit would be so unstable that at some point it might break down entirely and Nemesis would leave the Sun's influence and become a wandering lone star. There were also questions raised about the number of comets that Nemesis would send hurtling into the Solar System. The original model suggested that as Nemesis passed through the Oort cloud, millions of comets would be sent pummelling towards the Solar System. Yet, even taking into account the loss of craters as part of continuing geological processes, there didn't seem to be enough craters on our planet. Even though only a fraction of the meteors released by Nemesis would impact the Earth, we would expect to see more craters. Worse, a detailed statistical reanalysis of the timing of the impacts revealed that there was no match between the age of the larger craters and the mass extinctions, and no rhythmic pattern at all. In 2011, a newsletter from the German-based Max-Planck Society loudly announced: 'Nemesis is a myth.'

MASS EXTINCTIONS AND THE EARTH SYSTEM

So where does that leave us? Is there a general model of mass extinc-tions, something that unifies them all regardless of the details of the cause? Yes, there is – the Earth System. Remember that the Earth System treats the Earth holistically, seeing it as a series of physical reservoirs linked by complex feedback loops that cycle energy and essential elements around the system. These feedback loops allow the system to maintain the Earth's environment in a state where life can flourish. The biosphere is one of these reservoirs and should be seen as an active player in the system. You can think of it as life playing a leading role in ensuring its own existence.

If a severe environmental stress is applied to the Earth System, because the feedback loops link all the reservoirs, its effects will eventually be felt by the entire system. The stress may be severe enough to begin to shift the equilibrium, perhaps towards environmental conditions that are less suitable for life. This is the sort of process that would occur during the build-up to a mass extinction. It's the sort of long-term stress that we are applying to the system today through our high levels of carbon dioxide emissions and habitat change. Because the equilibrium is moving away from a state that is optimal for life, the rate of extinction will start to rise. The system will attempt bring itself back to something close to the original state of equilibrium. And if whatever is causing the environmental stresses is removed (say, the eruption ceases), this is likely.

A mass extinction occurs when the environmental stresses rise to the point where they can be considered a forcing and push the system past a tipping point, triggering a shift to a new equilibrium state – quite possibly one where the environmental conditions can't sustain the existing level of diversity. As a result, the already damaged ecosystem collapses, and we have a mass extinction. This shift in equilibrium state has not proved to be permanent. We will see that in the case of the ancient mass extinctions, recovery starts when the environmental forcings are removed. Once this happens, the level of extinction starts to fall, and evolution will begin to produce new species to fill the ecological gaps vacated by those species that went extinct. As the level of biodiversity rises, the ecosystem will recover, eventually allowing the natural feedback loops, crucial for the operation of the Earth System, to begin to operate at full efficiency. Because the Earth System has maintained the Earth's environmental conditions at more or less the same level for at least the last 550 million years, we can be reasonably confident that it will return the system to something like the pre-extinction planetary conditions. It may take thousands if not millions of years, but a new, fully functioning ecosystem will be restored. On the scale of deep time, this recovery may be rapid; but on a human time scale, it clearly isn't.

The Earth System provides a general model for all mass extinctions – environmental forcings cause a shift in its environmental equilibrium position causing a mass extinction. What I want to address in the next chapter is: what are the drivers of these environmental forcings? To put it more simply – what causes a mass extinction? We have already seen that in the case of the end-Cretaceous event a meteorite is certainly part of the story; but there are other potential drivers we need to discuss.

6 Causes of the End-Permian and End-Cretaceous Extinction Events

INTRODUCTION

Scientific reputations have been made and lost in extremely robust partisan discussions trying to definitively decide what causes a mass extinction. Michael Benton, in his book *When Life Nearly Died*, lists 100 possible causes for the extinction of the dinosaurs culled from literature published between 1920 and 1990. Suggestions include slipped vertebral disks, AIDS, development of psychotic suicidal tendencies (remember these were all real suggestions), gigantism, outcompeted by caterpillars, climate change (things got too hot – or too cold), and poisoning by toxic substances erupted from volcanoes. My personal favourite is the suggestion that they died of chronic constipation when angiosperms (flowering plants) replaced the ferns which had provided the roughage that kept them regular. Apart from the fact that some of the suggestions on Benton's list appear simply too outlandish to be true, the major problem with most, if not all, is that they focused on the question 'What killed the dinosaurs?' But that is not enough. Far more (than just the dinosaurs) of the global biota – both on land and in the sea – went extinct during the end-Cretaceous event. Earlier, we saw estimates suggesting that between 70% and 76% (Table 5.1) of all marine species on Earth went extinct at the same time as the dinosaurs. Constipated dinosaurs cannot, under any circumstances, be seen as a reason for all these other life forms going extinct.

So: the end-Cretaceous extinction event involved far more than just the dinosaurs. It was the result of the collapse of the entire global ecosystem, or at least a significant part of it, triggered by an equilibrium shift in the Earth System. As I discussed in the previous chapter, this is true for each and every mass extinction in the history of the

Earth: they are all caused by the collapse of at least a major portion of the global ecosystem, which followed a major forcing being applied to the Earth System. The question we should be asking about the end-Cretaceous event isn't 'What killed the dinosaurs?' but rather 'What caused the ecosystem to collapse?' To put it another way: what forcings triggered the shift in the Earth System that resulted in the collapse? The same question can be asked of all mass extinction events. In the case of the end-Cretaceous event, the Alvarez impact model – discussed in the last chapter and covered in more detail below – suggests that the trigger was a meteorite impact. Does it then follow that every extinction must have caused by a meteorite? No, it doesn't. Despite huge efforts by teams of scientists, the end-Cretaceous is the only mass extinction that is clearly linked to a meteorite impact. And as I will argue below, even the impact of a meteorite may not have been enough to wipe out over 70% of marine species. However, there is one geological phenomenon that appears to be involved in many of the major mass extinctions that have taken place over the past 500 million years. Most mass extinctions in that time interval coincide with massive volcanic eruptions which result in the emplacement of what is known as a Large Igneous Province or LIP.

LIPs represent some of the biggest volcanic events ever seen on the planet. During their emplacement, massive lava flows often covered millions of square kilometres of the Earth's surface to a depth of several hundred metres. They released vast quantities of volcanic gases into the atmosphere (in particular carbon dioxide and sulphur dioxide) which must have had a significant effect on the global environment. LIPs are thought to be a result of magma upwelling as a plume from deep in the Earth's mantle.

The close linkage in time between mass extinctions and LIPs[1] is illustrated in Figure 6.1. The Big Five mass extinctions are marked as

---

[1] I was once at a seminar where Walter Alvarez himself pointed out that the only possible cause that could link all mass extinctions was igneous activity. He was almost certainly only talking about the Big Five.

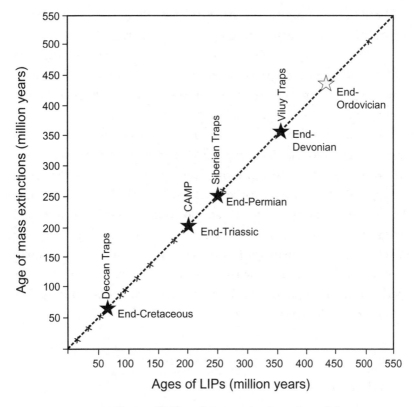

FIGURE 6.1 Timing of LIPs and mass extinctions. Ages of the mass extinctions are on the vertical axis, ages of the LIPs on the horizontal. Stars are the Big Five mass extinctions, crosses are other mass extinctions that are associated with the emplacement of LIPs. See text for full explanation. Figure adapted with permission from Keller et al. (2012).

stars. Most of them are dark coloured, but the end-Ordovician event is marked by a white star. This is to indicate that it is the only member of the Big Five that doesn't have a LIP associated with it. The black crosses that are scattered along the line are other mass extinction events that are associated with a LIP. The Deccan Traps are in India, and the Siberian and Viluy Traps are in the Russian Republic. CAMP is short for Central Atlantic Magmatic Province. Remnants of this LIP are today found in the south of North America and the northern parts of both South America and Africa. Before these continents drifted apart, all these remnants made up a single massive LIP.

The scale on both axes is exactly the same. Because the ages of the various mass extinctions can be matched exactly with the ages of the LIPs, the stars and crosses lie on the same straight line. This is strong evidence that there is a link between the two.

Figure 6.1 demonstrates that LIPs are implicated in most of the mass extinctions that have occurred over the past 400 million years. However, as we will see in some cases, they may not be the entire story, and an additional trigger may be needed. Rather than going through each mass extinction and looking at possible causes and triggers (there are many excellent books that do that), here I'm going to concentrate on just two. Firstly, we will look at the end-Permian event, the biggest of all mass extinctions. Then we will look at perhaps the most controversial of all mass extinctions, the one that killed off the dinosaurs – the end-Cretaceous event.

THE END-PERMIAN MASS EXTINCTION

The 95% Effect

It's been called the 'mother of all mass extinctions', and the name isn't surprising when you remember that the estimated species extinction rate among marine species was about 95% (Table 5.1). It's hard to imagine what the oceans would look like if 95% of all species were suddenly removed, so let's try to make it clearer.

Today's coral reefs are global hotspots of biodiversity. Many years ago, I was lucky enough to do some snorkelling around the reefs of the Bahamas, and they are astounding. In shallow lagoons that lap the Bahamian islands, millions of tiny coral animals build up carbonate mounds called patch reefs. In places, they grow so large that they almost break the water surface. The shallow lagoon floor surrounding the patch reefs is littered with rubble where jagged pieces of broken coral have fallen down the reef's sides. But patch reefs make up only a small proportion of Bahamian reefs. Most reefs form as long sinuous bodies fringing the lagoons, demarcating a line between the relatively shallow lagoons on one side and much deeper water on the other. Here, coral debris can tumble down many hundreds of metres before coming to rest on the sea floor.

All modern hard corals belong to a single taxonomic order, the Scleractinia. In the Bahamas, this order has diversified into over fifty different species. The most conspicuous of the corals are the elkhorns, which take their name from their resemblance to elk antlers. They build massive coral heads in the shallower parts of the reef, where their branching arms strain towards the sunlight. I visited the reefs in the early 1980s and remember the elkhorn corals well. Sadly, if I visited today the picture would be very different. Like many species of corals worldwide, elkhorns are under severe threat, largely as a result of climate-change-induced bleaching. They are now absent on many Bahamian reefs.

Around the corals swim more than 600 species of fish, and innumerable species of invertebrate animals find food and shelter between the coral heads. The level of biodiversity present in the Bahamian reefs is the highest recorded in the North Atlantic region. Yet even that level of biodiversity pales in comparison to the reefs of the Indo-Pacific, such as the Australian Great Barrier Reef.

There were great reefs present in the Permian too, and they are now re-exposed in limestones in many parts of the world. One of the most spectacular is the Capitan Reef in the Guadalupe Mountains that straddle the border between Texas and New Mexico in the United States. Named after a prominent peak in the range – El Capitan – the reef almost completely encircles an area of what was once an ancient sea called the Delaware Basin: an ancient fringing reef some 720 kilometres (about 450 miles) long. Today, thanks to uplift and erosion, whole sections of the reef have been exhumed, and the former topography of the ancient reef has been revealed.

The mountains that ring the Delaware Basin are topped by a thick, vertically sided cap of light-coloured limestone. This is the main reef itself, built by millions and millions of tiny coral polyps. Below the capping reef, towards the centre of the Delaware Basin, the slope shallows. The rocks here are largely composed of rubble: parts of the reef that broke away from the growing reef and fell towards the basin floor. At the bottom of the slope, you reach the ancient sea floor itself. Where once ancient organisms crawled across the muddy sediment, roads now carry a steady flow of traffic. On the other side of the mountains – behind the ancient reef – are flat-lying layers of rocks composed largely of muds and fine sands: these are common deposits in shallow back-reef lagoons.

Like modern reefs, these ancient Permian reefs were hotspots of biodiversity, although there were some significant differences: in the Permian, reefs were built by two orders of corals, Rugosa and Tabulata, and sponges also played a more significant part in the building of reefs than they do today. Like today's reefs, many species of fish and invertebrates lived in and around the coral heads. Despite this high level of biodiversity, not one coral reef anywhere in the world survived the end-Permian mass extinction. Every last rugose and tabulate coral that was living prior to the event – together with all the animals that relied on them for food and shelter – was wiped out. Paul Wignall, who knows more than most about the end-Permian event, suggests that the entire extinction event, which includes the reefs, took about 200,000 years: a geological instant.[2] For at least 27 million years following the mass extinction, no coral reefs grew on Earth. Anywhere. It wasn't until sometime about halfway through the Triassic that modern Scleractinian forms appeared, and the process of building new reefs could begin again.

The Environment at the End of the Permian

At the end of the Permian (and indeed all the way through to the early Jurassic), all the continents were assembled in a single massive

---

[2] The work of Paul Wignall forms the basis of much of the discussion of the end-Permian event.

supercontinent, which stretched from pole to pole, leaving most of the planet covered with water. This supercontinent is called Pangaea,[3] and climatically it was an unpleasant place. Modelling suggests that it was extremely hot, owing to very high carbon dioxide levels in the atmosphere (over six times the current level). Clearly there was something affecting the Earth System's ability to lower the levels of this greenhouse gas. There appear to be three factors operating together to cause this:

(1) Around each of today's continents is a wide area of shallow 'shelfal' seas (those covering the continental shelves) that are home to some of the animals that absorb carbon dioxide as they build their shells and eventually lock it away as limestone. Because Pangaea was only a single continent, the area of shallow marine conditions where these animals could live was smaller. This would inevitably lead to a reduction in the numbers of shell-building animals and hence the amount of carbon dioxide that can be absorbed.

(2) The major organisms that absorb carbon dioxide in the deeper parts of today's oceans are microscopic algae called coccolithospheres. They produce tiny calcium carbonate plates called coccoliths. Deposits of billions of these tiny plates form chalk, the rock that makes up the White Cliffs of Dover. Although minute, coccolithospheres are very efficient at locking away carbon, and because they can inhabit deep waters they wouldn't have been affected by the reduction in the area of shelfal seas that was caused by the assembly of Pangaea. Unfortunately, coccolithospheres didn't evolve until after Pangaea had broken up in the early Jurassic – so they are of no help in reducing the carbon dioxide levels during the Permian.

(3) Finally, and perhaps most importantly, while Pangaea existed there was a significant reduction in the level of chemical weathering which under normal circumstances would strip the gas from the atmosphere. Today, because of weather patterns, the central parts of large continents are often deserts or at least areas of intense aridity. As I discussed in Chapter 1, for chemical weathering to take place and strip carbon dioxide out of the

---

[3] The name Pangaea is derived from the Greek words *pan*, meaning all or entire and *Gaia* – the Earth Mother. We have met her before.

atmosphere, warm humid conditions are needed. So there is not much chemical weathering taking place in the centre of continents.

The same would apply to Pangaea. The centre of the continent would be arid and the level of chemical weathering extremely low. However, the huge size of the continent means that a huge area of the Earth's surface would be dry, and there would be very little chemical weathering anywhere on the planet, maintaining high levels of carbon dioxide in the atmosphere.

Taken together, the reduction of the area of shelfal seas, the lack of deep-water-living shell builders, and the reduction in the level of chemical weathering inhibited the Earth System's ability to control the level of carbon dioxide in the atmosphere. The carbon dioxide pumped out by volcanoes simply built up in the atmosphere, and because of its greenhouse properties, global temperature rose.

At the same time as the extinctions were occurring, strange things were happening in the Permian Ocean. We have seen already that oxygen is a prime requirement for life. Today, the surface waters of the ocean are well oxygenated, with abundant oxygen diffusing into the water from the atmosphere. Current and wave action ensures that this oxygen is well mixed into the upper water layers. In general, the level of oxygen decreases with depth, as organisms consume it and the slower deep-ocean circulation doesn't mix the water as well. There are also places in today's ocean called oxygen minimum zones, where, as their name suggests, the level of oxygen is very low. These are found in places where nutrient-rich deep water wells up to the surface, and an abundance of organisms, benefiting from the high nutrient levels, consume most of the available oxygen. Because any oxygen that organisms use in the upper water layers is constantly replenished from the atmosphere, oxygen minimum zones never usually come close to the surface. However, in the Permian, at the time of the extinction, it seems that the whole ocean from the north pole to the south pole, from the sea floor right up to the surface, became an oxygen minimum zone – all the ocean waters became anoxic. Because this includes the surface waters, it suggests that the supply of oxygen from the atmosphere was severely reduced, at the very least.

It's a bit harder to document exactly what happened on land, since once again we are limited by the fossil record. But there are places that provide a window into what went on. One region with a particularly good terrestrial fossil record across the Permian/Triassic boundary is in the Karoo region of southern Africa. Rocks from this region record the annihilation of many groups of vertebrates, with only a few species staggering through the event. There's a good record of insects, and it shows that up to 40% of families disappeared. Surprisingly, there seemed to have been little effect on organisms that live in fresh water.

Despite the fact that the fossil record is undoubtedly predisposed to preserving organisms with hard parts, we have a very good record of pollen and spores across the extinction event. This is because they are composed of a very unreactive material called sporopollenin which is chemically inert and easily preserved. Analysis of the Permian spores and pollen shows that immediately prior to the event, lush forests full of tall trees stretched from pole to pole. At the beginning of the event, these trees quickly disappear, to be replaced with a low shrubby vegetation. Coal is common in the late Permian – thick seams are found in many of the southern continents including Antarctica – but the production of coal seems to have stopped at the end of the Permian. There are no coal seams of early Triassic age anywhere on Earth. By the mid Triassic, thin and sporadic seams are starting to appear, but it's not until the late Triassic, some 20 million years after the mass extinction, that we start to see thick seams of coal in the record again. This interval has been named the 'Triassic coal gap'. Coals form in swamps and other areas of intense plant growth, so the coal gap suggests that these environments vanished soon after the end-Permian event started, and it took a long time for them to be re-established.

So, the environment both on land and in the ocean at the close of the Permian was unpleasant to say the least, and the ecosystem must have been significantly stressed – but not so much as to cause a mass extinction. For that to happen, the emplacement of a LIP is required.

THE SIBERIAN TRAPS

The end of the Permian was clearly not a good time to be alive. In fact, Paul Wignall's book detailing the end-Permian event is called *The Worst of Times*, and that seems to me to be a highly appropriate title. The book examines the 80 million years starting with a mass extinction that occurred just prior to the end-Permian event (the Capitanian extinction that I talked about earlier) and ends with another mass extinction in the early Jurassic. Here I'm only going to concentrate on Wignall's model for the end-Permian event, which links the extinctions with the emplacement of the Siberian Traps,[4] a huge LIP.

Today, the lavas produced during the eruption of the Siberian Traps cover about 2 million square kilometres of the northern and central part of the Russian Republic. They were erupted about 250 million years ago, so much has since been eroded away. Estimates of the original area range up to 7 million square kilometres: that's about the area of Australia. The lava wasn't spread thinly. It is possible that up to 5 million cubic kilometres of lava was extruded over the million years or so that they took to erupt. Scientists who have examined the Siberian Traps in detail suggest that their eruption was, at least at the beginning, extremely explosive, blasting debris and gases high into the stratosphere.

If the figures for the amount and distribution of the lavas seem mindboggling, here's some more for you. All volcanic eruptions will release gases into the atmosphere, dominantly carbon dioxide, but also halogen gases (gases involving fluorine, chlorine, bromine, and iodine atoms, for example), and hydrochloric acid (HCl). But gases bubbling directly from the lava may not have been the only source of carbon dioxide and halogen gases. Modelling suggests that as the hot molten magma pushed its way to the surface, it would generate

---

[4] The term 'Trap' is often associated with LIPs. It's the Dutch word for stairs and refers to the layer upon layer of lavas produced during its emplacement, which tend to look like stairs. We get the English word trapdoor from the same source.

still more of these gases by heating the crustal rocks it was passing through. One estimate suggests that this combination of sources released 170,000 gigatonnes (170×10^{12} tonnes) of carbon dioxide into the atmosphere, and possibly 17,000 gigatonnes of hydrochloric acid. I have a friend, Dave Frame, who is an atmospheric physicist. I showed him these estimates of the amount of carbon dioxide released, and he was shocked and inclined to disbelief, so I went away and confirmed the numbers (Paul Wignall is my source for these estimates). After a quick back-of-the-envelope calculation, Dave concluded that the 170,000 gigatonnes of carbon dioxide that may have been released over the million years of the eruption is roughly equivalent to 92 times the total amount of carbon dioxide that humans have added to the atmosphere since the industrial revolution. Look at the effects that we are having on the planet today and imagine the havoc that would be caused by the injection of that much carbon dioxide into the atmosphere.

Figure 6.2 sets out Wignall's model of how the Siberian Traps could trigger the greatest loss of life ever seen on Earth. Starting at the top, the two white boxes at the top of the figure show the eruption of the Siberian Traps and the subsequent release of huge volumes of gases both directly from the lava and from heating the crustal rocks. Below these boxes, the flow chart splits into two – the light grey boxes are events that lead to marine extinctions and the dark grey to terrestrial extinctions. Let's follow the light grey to start.

As we have seen, carbon dioxide is a very effective greenhouse gas. Simple physics shows that increasing the levels of carbon dioxide in an atmosphere leads to higher global temperatures. We are currently demonstrating this lesson in simple physics by increasing levels of this gas in today's atmosphere, resulting in a warming climate. This greenhouse effect has also been clearly demonstrated on other planets. For example, on Venus the atmosphere is almost entirely carbon dioxide, which has triggered a runaway greenhouse effect resulting in surface temperatures of over 450 degrees Celsius. The huge volumes of carbon dioxide released into the atmosphere

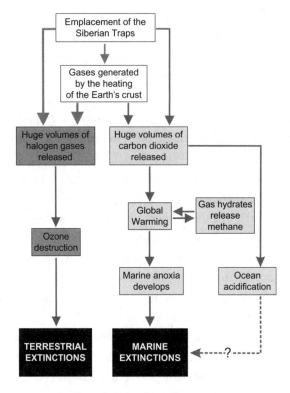

FIGURE 6.2 Flow chart for the end-Permian mass extinction. See text for details. Adapted with permission from Wignall (2005).

during the end-Permian eruption would result in something like the Venusian runaway greenhouse effect. (Note that I am not claiming that the temperature reached 450 degrees – otherwise nothing would have survived the extinction!)

To make matters worse, the greenhouse effect of the carbon dioxide was amplified. Today, in sediments at the bottom of the cold, dark, and deep ocean just off the margin of many continents is a compound called a gas hydrate. It consists of a frozen water 'cage' that traps a gas called methane. As long as it is cold, everything is fine, but if a gas hydrate warms and melts, it will release all the methane it contains. And it will release a lot of methane – 1 litre of hydrate produces 130 litres of methane. Methane is also a greenhouse gas – a much more effective one than carbon dioxide. However, it breaks down more quickly in the atmosphere, and so its greenhouse influence, although large, lasts for a much shorter time than carbon dioxide.

There are geochemical signals in late Permian sediments that suggest that in response to carbon-dioxide-induced global warming, gas hydrates gave up their methane, leading to a further increase in the greenhouse effect. This would have resulted in a feedback loop developing: melting of more gas hydrates leading to further increases in global temperature, leading to more melting of gas hydrates, and so on. There are some climate scientists who are very concerned that this is a possibility in today's warming climate. Perhaps of more concern is that the residence time of methane in the atmosphere is so short that it is difficult for our current predictive climate models to take it into account.

Back to the Permian. As the greenhouse effect really kicked in and the Earth System struggled to bring the level down, the oceans warmed, and this affected the ability of the near-surface waters to absorb and transport oxygen. Again, this is simple physics: as the temperature of water increases, its ability to absorb and carry oxygen decreases markedly. If the near-surface waters of the Permian ocean were unable to take up atmospheric oxygen, they would become anoxic. This would eventually stop the supply of oxygen to the deeper water, and the surface anoxia would eventually spread to the entire ocean, triggering the mass extinction of marine species.

This mass killing may have been enhanced (if that's the right word) by some of the carbon dioxide in the atmosphere dissolving into the sea water. This would cause the surface waters to become acidic (carbon dioxide + water = carbonic acid).[5] Today, acidification of the oceans is of great concern, as acidification may dissolve large numbers of the microorganisms that build calcite shells, triggering some

---

[5] To be fancier, the formal equation is: $CO_2 + H_2O \rightarrow H_2CO_3$.

marine extinctions. Whether or not the acidification of the Permian ocean contributed to the end-Permian mass extinction is uncertain, but it is certainly a possibility, and it is indicated by a dashed line on Figure 6.2. The extinction of these shell builders would have also caused climatic effects. In our discussion of the Earth System and the carbon cycle (Chapter 1), I showed how today we rely on those shell builders to strip out some of the excess carbon dioxide that we are putting into the atmosphere. The extinction of the creatures at the end of the Permian would have led to a further increase in the level of carbon dioxide in the atmosphere, making the greenhouse situation worse. Feedback loop upon feedback loop, it could have led to a runaway greenhouse effect.

There is, however, a problem with this model for the end-Permian marine extinctions: we could call it the '4% problem'. As Dave Frame has pointed out, the amount of carbon dioxide being injected into the atmosphere almost beggars belief. Why didn't we get a complete runaway greenhouse as we see on Venus? How did even 4% of marine species survive? There is no clear answer to this – but we know they did survive. Additional work will be needed to sort out this paradox.

The cause of the terrestrial extinctions is less clear. Follow the dark grey boxes of Figure 6.2 down and we can explore Wignall's suggestion. We have seen that as well as carbon dioxide, the eruption of the Siberian Traps would have released large amounts of halogen gases. These gases are known to damage the thin ozone layer that protects the Earth's surface from the Sun's damaging ultraviolet (UV) rays. In order to destroy the ozone layer, the gases have to be injected high into the stratosphere. Any lower, and they would be washed out of the atmosphere by rain before they could cause damage. All the evidence suggests that, at least in the initial stages, the eruption of the Siberian Traps was explosive enough to contaminate the stratosphere with halogen gases, resulting in a removal of the protective ozone layer.

The bombardment of the Permian vegetation with high-energy UV radiation could have caused the shift in the fossil record of plants and the disappearance of coal swamps that I talked about earlier. It would also explain why organisms under the protective blanket of fresh water survived. UV radiation can cause significant damage – but I do wonder if this is the whole story for the terrestrial animals. Perhaps there we are looking at a combination of causes:

- The direct effect of the ozone removal and the impact of UV radiation;
- The increase in temperature resulting from an enhanced greenhouse effect;
- The collapse of the ecosystem that results from the loss of much of the planet's vegetation.

Regardless of any uncertainty about the exact kill mechanisms, the link between the end-Permian extinctions and the emplacement of the Siberian Traps appears to be solid. Evidence is also accumulating that suggests that Wignall's model explaining the linkage between LIPs and mass extinctions applies to many ancient mass extinctions. Reports of increased levels of carbon dioxide, oceanic acidity, and anoxia associated with these events are appearing regularly (see the Further Reading section). But a question that we will come back to in more detail when I talk about the end-Cretaceous mass extinction is whether a mass extinction can be triggered by the emplacement of a LIP alone. In the case of the mass extinctions that occurred between the end of the Permian and Jurassic, Wignall argues that the environmental stress applied to the biota by the environmental conditions over this time interval predisposed it to mass extinctions. But what triggered the mass extinctions were the environmental changes wrought by the emplacement of a LIP. This idea that each mass extinction is a two-part event – a longer-term stress combined with a short sharp shock – will dominate our discussion of the end-Cretaceous mass extinction,

THE END-CRETACEOUS MASS EXTINCTION

A Dramatic Introduction

Before we move onto more serious matters, let me indulge myself by going through how I began my lecture on the end-Cretaceous event to my first-year geology class. I started (somewhat melodramatically) by turning off most of the lecture theatre lights, leaving only what I liked to call 'mood lighting'. In the semi-darkness, I presented the following scenario:

Imagine you have drifted back through time to a point 66 million years ago. You are on a beach looking south across the future Gulf of Mexico. To get here you have walked across the plains of what one day will be southern Texas. On your walk you passed herds of dinosaurs, mainly Triceratops, as they wandered majestically across the plains in search of better grazing. It's June, mid-Summer, and you stop briefly by a pond containing flowering water lilies, but disappointingly the flowers have passed their prime.

A sharp noise from under a nearby tree alerts you to the presence of a large meat-eating dinosaur. You have disturbed him (or her – it's hard to tell at this distance, or any distance really) from its rest after a hearty meal of Triceratops. It looks at you, decides you're not dangerous (or edible), preens its feathers, then goes back to sleep.

Now, as you gaze out to sea, you notice a sudden increase in the intensity of the light, looking up – there's a second sun in the sky! Quickly the light's intensity increases to a point that is not normally seen outside of Stephen Spielberg's early movies. Then there's a huge explosion, and a shockwave that knocks you to the ground. The sound is all-enveloping; it's as if the air itself is being torn apart. In the far distance you can see dinosaurs stampeding in panic. The second sun is getting bigger, it's dominating the sky, and the heat all around you is intense. As you watch, the second sun rushes to the horizon and sets to the south of you.

There is silence – except for the trumpeting of panicked dinosaurs. Soon you notice a vast cloud of dust rising from where the imposter sun set. The dust column grows and the cloud rushes towards you. It starts to block out the real sun. Premature darkness sets in.

In the howling dust storm, large chunks of white-hot rocks start to fall around you, triggering wildfires. Smoke and soot from the blazes start to rise, joining the swirling, suffocating, dust. Eventually the cloud of dust and soot blocks out the sun, and complete darkness nears. It starts to rain, but it's not water you can feel on your skin, but acid. By the last light of the fading sun, you see approaching across the gulf a huge tsunami, perhaps 150 metres high, rushing towards the coast.

It was great fun (at least, I enjoyed it, and the first years didn't complain). The scenario is a very much fictionalised account dramatising the meteorite impact model. But it's not all science fantasy: I could justify all my claims from the scientific literature. This includes the water lilies – see the paper by Wolfe in the Further Reading section. Its geographic setting is carefully chosen to reflect the suggested location of the meteorite crater in present-day Yucatan. I think (I hope) that I have all the events in the right order, although I have expanded some of the action and telescoped the rest for the sake of drama. I changed it each time I presented it, adding and subtracting elements as new information came to hand. In the first draft of this book, I removed any references to global wildfires. At the time, the idea of impact-induced global wildfires appeared to have been largely rejected. However, I've had to put them back. The latest research suggests that wildfires were an important part of the extinction event – although they seem likely to have been limited to within a few thousand kilometres of the impact site. Sadly, owing to changes in the curriculum, I don't get the chance to talk about mass extinctions to first years anymore. Perhaps I could sneak them back in,[6] along with insisting they read this book!

A MORE FORMAL INTRODUCTION

The end-Cretaceous extinction occurs synchronously with both the emplacement of a LIP, the Deccan Traps, and a meteorite impact. The

---

[6] I do continue to give a lecture on the end-Permian mass extinction to the second years. I start that lecture by playing the theme to *Apocalypse Now*. It's probably too late for me to grow up.

emplacement of the Deccan Traps in India resulted in lava covering an area of about 1.5 million square kilometres, to a depth of about 2,000 metres. As with the Siberian Traps, it was accompanied by the release of huge amounts of carbon dioxide and other gases. One estimate suggests that 425 ± 180 gigatonnes of carbon dioxide and 325 ± 130 gigatonnes of sulphur was introduced into the atmosphere over the million or so years it took to emplace the Deccan Traps. This would produce similar conditions to those that existed at the end of the Permian and could, conceivably, have caused the end-Cretaceous extinctions. There are, however, significant numbers of scientists who deny that the Deccan Traps had anything to do with the extinctions. This extremely forceful group insist that the meteorite suggested by the Alvarez team was more than enough to cause the mass extinction on its own. I don't think that it's an overstatement to say that this clash of ideas has resulted in controversy and a great deal of bitterness and division, which, even as I write, shows little sign of going away. It has not been an edifying exchange of ideas. So, at the risk of losing friends, let's look at the end of the Cretaceous.

I have already noted that most people's attention when thinking about what happened is concentrated on the extinction of the dinosaurs, but we mustn't forget that they are only a part of the story. We have already seen that in the sea about 75% of species died. All the marine reptiles, mosasaurs, plesiosaurs, the dolphin-like ichthyosaurs all disappeared, and so did the coil-shelled ammonites, relations to today's nautilus. Micro-invertebrates, such as the foraminifera (related to the single-celled amoeba familiar from biology classes), and marine algae were all decimated.

The extinction cut great swathes through terrestrial communities. As well as the dinosaurs, pterosaurs – flying reptiles – also perished. There are suggestions that, like in the end-Permian event, insects were also badly affected. Mammals made it through, but not without severe losses. The effect on plants was complex. In some locations, particularly in North America, there was a significant number of extinctions of species, with some estimates suggesting a

57% kill rate. Further south, in places like Aotearoa/New Zealand, there doesn't seem to have been a major plant die-off. Here the boundary is marked by a sharp, but temporary, spike in the number of ferns present. This fern spike has also been recorded from sites in the Northern Hemisphere. For a brief period, ferns seem to have become the major plant group on Earth. Ferns are ecological pioneers, the first plant group to return to an area following the clearing of the previous community. The presence of the fern spike suggests that it is a response to a significant clearing (if not extinction) of plant communities, as a result of whatever happened at the end of the Cretaceous. We'll come back to this in the next chapter.

MORE ON THE METEORITE

I noted at the beginning of this chapter that the first piece of physical evidence suggesting that a meteorite hit the planet at the end of the Cretaceous was a massive increase in the level of the element iridium found in the boundary clay from sections like the one in Gubbio, Italy. Since then, much more evidence supporting the reality of the impact has been accumulated. Additional analysis of the boundary clay from sites around the world shows that as well as the iridium, it contains the chemical signature of the pulverised debris of a meteorite. It also contains crystals of shocked quartz. It's hard to shock quartz: you have to deform a very stable crystal structure, which requires high temperatures and pressures. Before its recognition in the boundary clay, shocked quartz has only been found at atomic test sites and in meteorite impact craters. The debris from the meteorite – which includes the iridium – together with remains of the rocks that the meteorite hit was all blasted into the stratosphere by the impact. This mix of material, a combination of meteorite debris and pulverised target rocks, settled out globally as the boundary clay (and eventually found its way underneath the fingernails of many geologists).

Based on their analysis of the clay layer from sites around the world, the Alvarez team estimated that the meteorite would have been about 10 kilometres in diameter,[7] and it had struck the Earth with the same force as detonating the entire globe's stockpile of nuclear weapons 10,000 times. The impact produced a huge crater: it is 200 kilometres in diameter and is located near Chicxulub on the Yucatan peninsula. You can't visit the crater, as it is now buried under many metres of younger sediments, but the crater's edge can be seen using remote sensing techniques as a ring, arcing parallel to Yucatan's northern coast, then swinging out to sea for many kilometres before returning to the coast. Aside from the fact that the formation of the crater and the extinctions occurred at exactly the same time, additional evidence that Chicxulub was ground zero for the meteorite comes from Montana in the United States. There, tiny crystals of a mineral called zircon have been recovered from the boundary clay. The Montana crystals have been chemically fingerprinted and found to exactly match zircons that could only have come from the Chicxulub crater.

There is abundant geological evidence documenting the effects of the meteorite. For example, the impact caused huge tsunamis to spread across what is now the Gulf of Mexico.[8] These massive waves left tell-tale sedimentary deposits that we can now find along the coasts of Latin America, Mexico, and the southern United States. Also in the United States, a geological section from North Dakota records a 140-centimetre-thick layer of sediment sitting just below an iridium-rich boundary clay. This layer of sediment, which contains a number of dead fish, has been interpreted as a surge in a river triggered by the arrival of a seismic wave about an hour after impact. The debris to create the boundary clay would have fallen from the sky shortly afterwards.

The crater itself has been the focus of considerable study. Geophysical imaging techniques have mapped out its size and shape.

---

[7] In some papers, it seems to have been upgraded to between 12 and 15 kilometres.

[8] One attempt at modelling the impact suggested that the wave could have been as high as 1.5 kilometres. See Kornei (2018) in the Further Reading section.

Images of the crater based on gravity measurements clearly show a double ring structure, with a gap in the northeast section of the outer ring. The results of a recent project that drilled into the crater and sampled the impact rocks have provided a timetable of events that occurred during the day following the end of the Cretaceous.[9] On impact, the meteorite excavated a massive, but transient, crater some 30 kilometres deep and perhaps 100 kilometres wide, blasting out the target rock as dust and ejecta. Within minutes, the crater collapsed back, forming a ring of hills at its centre. These hills also rapidly collapsed, forming the final crater. Thirty minutes later, the sea flooded back in, filling the crater. In less than a day, the tsunamis triggered by the initial impact returned after being reflected from the surrounding coasts. The returning waves carried back material stripped from the surrounding land areas. This debris was mixed with fragments of the target rocks and deposited in the crater. In the face of all this evidence, it is clear that the meteorite is real, and its synchronicity with the extinction is extremely unlikely to be a coincidence – so how did a meteorite cause the ecosystem collapse?

The original Alvarez model suggested that the primary reason was the blocking out of the sunlight by all the dust and debris blasted into the stratosphere by the impact. Recent work stresses the importance of the addition of soot from the wildfires to the mix. It appears that dust alone would not have been enough to completely block out the sunlight: soot is needed to do the job properly. Once the Sun was hidden by the cloud of dust and soot, the lack of sunlight resulted in the extinction of the plants that rely on photosynthesis to live. This would be followed by the extinction of plant-eating organisms, and finally the top predators – the carnivores that consume the plant eaters. There are other killing mechanisms associated with the impact model. For example, as the meteor passed through the

---

[9] This information is based on a drilling project carried out by the International Ocean Discovery Programme and published in a paper titled *The first day of the Cenozoic.* Great title – see Gulick and others (2019) in the Further Reading section.

Earth's atmosphere, it would have compressed the air in front of it. The compression of the air is thought to have been so intense that it forced the oxygen and nitrogen in the atmosphere to combine, producing nitric acid. This ultimately rained out, acidifying the ocean and enhancing the extinctions there.

AN ACCIDENT WAITING TO HAPPEN

So, a meteorite impact could trigger an ecosystem collapse and the end-Cretaceous mass extinction. But on its own, would it be enough? Amid all the excitement that surrounded the meteorite model, there were naysayers who were concerned that the model completely ignored the synchronous emplacement of the Deccan Traps. These workers believed that a volcanic eruption of the scale of the Deccan Traps would have had an impact on the environment, and contributed, at least in part, to the level of extinction that occurred at the end of the Cretaceous.

Leading critics of the meteorite-only model were palaeontologists who examined the fine detail of the extinctions across the Cretaceous–Paleogene boundary, and thought they could see a structure to them. By that I mean that they could identify some species that went extinct prior to the main extinction event, and some other species that flourished across the boundary. The extinction event itself was followed by a stepwise appearance of new species. This pattern of extinctions presents a problem. If a meteorite was the sole cause of the end-Cretaceous, there cannot have been any precursor extinctions. A meteorite impact is a *here today and dead tomorrow* deal: instantaneous extinction. Animals living near the end of the Cretaceous didn't wake up one morning and think, 'Oh, a meteorite is coming – I'd better go extinct now and avoid the rush.' So, is there anything in this notion of a structure to the end-Cretaceous extinctions, or should we dismiss it entirely?

The idea that there is a structure to the late Cretaceous extinctions and that this structure is telling us something fundamental about the cause of the event has been around since well before the impact model. The data largely come from the detailed analysis of records of microfossils called foraminifera in samples from across boundary sections – in particular, sections recovered from cores taken from deep-sea drilling. Foraminifera are single-celled organisms related to the Amoeba, and they secrete a shell or test the size of a sand grain.[10] These tiny animals are abundant in today's ocean and are also extremely common fossils. They have either a planktonic lifestyle, floating in the surface waters of the ocean, or a benthonic lifestyle, living on or in the sediments on the sea floor. Importantly, because foraminifera are abundant, easily preserved, and minute, they can be collected in large numbers from relatively small samples. These factors mean that we can produce highly detailed records of their abundance through time – perfect for documenting what happened across the Cretaceous–Paleogene boundary.

Figure 6.3 is part of a record of the foraminiferal abundance changes from an Ocean Drilling Programme (ODP) site on the Kerguelen Plateau in the Indian Ocean. Drawn by Gerta Keller, this figure's vertical axis is depth in centimetres – this is a high-resolution record, so samples were taken at very small intervals. The core has been dated, and the Cretaceous–Paleogene boundary, the moment of meteorite impact, is marked by the dashed grey line. The bars and polygons show the varying abundances of different species through the section. For simplicity, I've left off the species names. The dark grey bars and polygons are Cretaceous species. The species shown as simple black bars all go extinct, in a stepwise fashion, *before* the boundary. The species represented by the polygons in general do trickle across the boundary.[11]

---

[10] Unusually, some tests can be bigger. One specialist group called, unsurprisingly, larger foraminifera, produces tests several centimetres in diameter: the largest shell produced by a single-celled organism.

[11] The appearance of Cretaceous foraminifera in younger sediments is controversial. Many people consider it an artefact of the record called reworking – the transfer of older fossils into younger sediment as a result of erosion and redeposition.

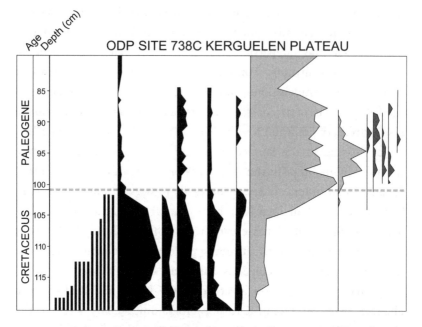

FIGURE 6.3 Foraminiferal abundances from deep-sea core collected on the Kerguelen Plateau. The age and depth of the core are shown down the left-hand side, and the Cretaceous–Paleogene boundary is marked by the dashed line. The bars and polygons show the distribution of foraminiferal species across the boundary – more detail in the text. Redrawn with permission from Keller (1993).

However, the width of the polygons represents how abundant each species is: the wider the polygon, the more abundant the species. Notice how the abundance of every species in this group drops sharply as they approach the boundary. The extinctions prior to the boundary and drop in the abundance near to it suggests that something was going on prior to impact, something that stressed the environment to the point where some species had trouble surviving.

The two species represented by the light grey polygons are often called 'disaster species' or 'survivor species'. Whatever was causing the extinctions and abundance drop amongst the Cretaceous species didn't upset these two at all – they appear to be thoroughly enjoying it, blooming across the boundary. The remaining species coloured medium grey are the Paleogene species that evolve after the extinction event, gradually replacing the Cretaceous species that went extinct.

What we are seeing on the Kerguelen Plateau – and the same pattern has been documented elsewhere with different fossil groups – suggests that there is turnover of species across the boundary that was not *instantaneous*. Rather, it started before the event itself. And although, on a human time scale, this turnover took some length of time, it was very rapid in terms of deep time. The pattern of extinction and replacement shown on the Kerguelen Plateau suggests that the end-Cretaceous event was, at least in part, the result of a longer-term environmental stress that started before the main event itself.

An obvious suspect for causing this stress is the emplacement of the Deccan Traps. In the discussion of the Permian/Triassic event, we saw the damage that the injection of all the volcanic gases associated with the emplacement of a LIP could do to an ecosystem. Enhanced greenhouse effect, ocean acidification, and the removal of the ozone layer all played a part in the end-Permian extinction event, and all were a result of the huge amount of volcanic gas generated during emplacement of the Siberian Traps. The Deccan Traps are smaller, so the amount of gas produced would have been smaller – but it would still have been large enough to inflict significant damage to the environment. Nevertheless, as I noted above, there are a significant number of scientists who deny the Deccan Traps any role in the extinctions; they attribute all the extinctions to the meteorite impact. There appear to be three ideas behind this denial:

(1) *There is overwhelming physical evidence for a meteorite impact.*

There is no denying this. Evidence for the impact is indeed overwhelming, and I have discussed some of it above. There are some who quibble about the exact age of the impact, doubting that it coincides exactly with the extinctions – but they are in the minority. The vast majority of scientists accept that the impact is synchronous with the extinctions. But so are the Deccan Traps. It seems perverse to deny the volcanic eruptions any role in the extinctions simply because we have strong evidence for a meteorite.

(2) *The species overturn isn't real – it's a quirk of statistics.*

This is the big one. If the pattern of species extinctions such as we saw on the Kerguelen Plateau isn't real, there is no evidence of longer-term ecological stress, and the Deccan Traps could be irrelevant. The issue here is to do with the quality of the fossil record (again) and how we sample it[12] – it's called the Signor–Lipps effect, named after the scientists who first recognised it. It's related to the problem we saw earlier when we discussed the difficulties encountered when we try to document the first appearance and final extinction of a species in the rock record (Figure 2.2).

Figure 6.4 is an attempt to show how the Signor-Lipps effect can change our perception of a mass extinction. Imagine four species (labelled 1,2, 3, and 4 on the figure) that all go extinct at exactly the same time – sort of a mini mass extinction. The upper chart of Figure 6.4a shows what a palaeontologist would record if we had a continuous sampling and we could collect the fossil of every individual of each species. In other words, this is what a perfect record would look like. Species range, without break, from the bottom of the section right through to where they go extinct. The lower chart of Figure 6.4a illustrates the pattern of extinction that we would derive from this perfect record. No extinctions until we reach the extinction event – then they all disappear at exactly the same time.

Unfortunately, we don't have a perfect fossil record, so let's look at what happens when we introduce some reality in the form of a patchy record and discontinuous sampling to the scenario. This is what is attempted in Figure 6.4b. As we go through this part, remember that Figure 6.4a represents reality – all four species really do go extinct simultaneously. The upper chart of Figure 6.4b shows the samples as a series of vertical grey dashed lines, each placed at the

---

[12] Which by now must be becoming a familiar refrain.

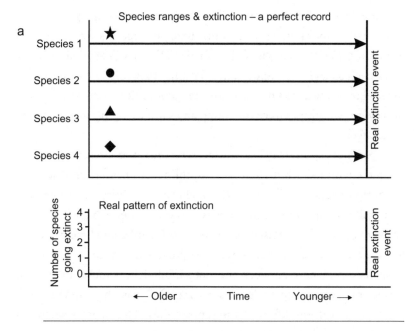

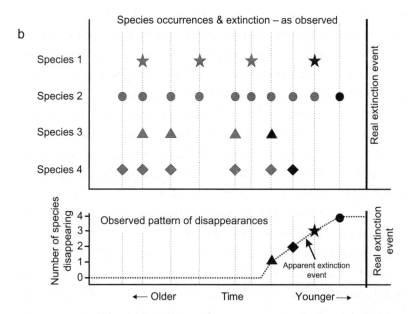

FIGURE 6.4 The Signor–Lipps effect in action. **a**, The species data and the pattern of extinction we would expect to see in a perfect world. **b**, The effect of a patchy fossil record and discontinuous sampling.

position from which it was collected. Only ten samples were collected, so sampling is far from continuous. If a species was recovered
from a sample, it's shown with a shape. The shapes shaded in black
mark each species' last appearance datum – their LADs. Note that all
the species seem to disappear from the record prior to the real extinction event (we simply have not found any samples of the species closer
to the event).

Species 2 (circles) has the best record. It is present in every
sample examined – but there is a sampling gap before we reach the
extinction event itself. I have discussed the difficulty in recognising
first and last appearances in the fossil record earlier (Chapter 2). That
problem applies here. Based on the data in Figure 6.4b alone, we can't
tell exactly where species 2 went extinct. All we know is that it
happens somewhere between the last sample it was found in and the
extinction event proper. Species 3 (triangles) has the worst record,
being present in only four of the ten samples. This species seems to
disappear well before the extinction event. In this case, the gap
between the LAD and the real extinction event is due to a combination of a patchy record and discontinuous sampling. The lower chart
of Figure 6.4b plots the disappearance of all four species against time.
I have just transferred each species' LAD down to this chart. The
dashed line, derived from the LADs of each species, represents the
changes in the apparent level of extinction over time. One species
after another seems to disappear as we approach the real extinction
event: first the triangles, followed by the diamonds, stars, and finally
the circles. Taken at face value, it looks as if species are gradually
dying out before the actual extinction event.

A comparison of the lower charts from Figure 6.4a and b clearly
demonstrates the Signor–Lipps effect. The sharp mass extinction that
is documented in Figure 6.4a – which is what really happens – is
smeared out in Figure 6.4b. The dashed line does not reflect the reality
of a sharp extinction event at all, and it looks like a gradual overturn,
like the one shown in Figure 6.3. The Signor–Lipps effect, a combination of the failure of the fossil record and the limitations produced by

discontinuous sampling, can mask a sharp mass extinction, making it appear to be a gradual species overturn. This has led to the claim by impact-only proponents that any data showing a gradual overturn of species across the end-Cretaceous event – such as the results from the Kerguelen Plateau (Figure 6.3) – are simply a statistical artefact.

Is this the end – we can't trust any data that show any sort of gradual change across a mass extinction event? Well, for some animal groups the answer will be yes. Any fossil group that has a very discontinuous record or can only be collected from a section that is poorly sampled will be affected, often badly. However, data like those from the Kerguelen Plateau usually come from studies that use very high-resolution data sets. Yes, there are gaps between the samples, but they are very small. In addition, they use fossil groups with an excellent record, such as the foraminifera. Like many other microfossils, many species of foraminifera usually produce almost complete fossil records. As a bonus, because they are small, they can be collected from sections like drill cores which offer an almost complete sampling opportunity, since the sediments that form the cores fell to the ocean floor pretty much continuously. Many microfossils species are like species 2 in Figure 6.4b – often present in all the samples we collect. If we can make the gap between the final sample and the event itself very small, we will get close to a perfect record. Under these circumstances, I would argue that even if we can't eliminate it entirely, the Signor–Lipps effect is minimised.

But that only covers the species that went extinct prior to the boundary, represented by simple bars on Figure 6.3. How does the Signor–Lipps effect deal with species where their abundance is decreasing as we approach the boundary, those represented by the dark polygons in Figure 6.3? The short answer is that it doesn't. The Signor–Lipps effect is only applicable when we are dealing with presence and absence data like Figure 6.4b. It can't cause the striking decline in abundance we see leading up to the boundary. This strongly suggests that the overturn of species recorded from sections like the one on the Kerguelen Plateau, and from many others around the world, has some basis in reality (rather than being purely statistical) and reflects a true increase in environmental stress leading up to the main extinction event, so it must be included in any explanation for what happened at the end of the Cretaceous.

(3) *There is no physical evidence to suggest that the Deccan Traps had any effect on the environment.*

Having dismissed the possibility of a species turnover at the end of the Cretaceous as a quirk of statistics, impact-only proponents have claimed that there is no physical evidence in the record to support the reality of longer-term environmental stress leading up to the boundary event. The sort of evidence we would be looking for would be similar to that recorded at the end of the Permian: evidence of long-term anoxia in the ocean, geochemical signals for clathrate release, and the like. For the end-Cretaceous event, this sort of evidence is starting to accumulate. Although it's not yet as strong as we saw at the end of the Permian, it is there. Recent studies from disparate areas of the globe demonstrate the sort of data that support a longer-term build-up to the mass extinction event. Here is some of that evidence.

The Boundary in Meghalaya

Let's first go to northeastern India – an area called Meghalaya, 1,000 kilometres or so from the Deccan Traps. Here, a team headed by Brian Gertsch carried out a detailed palaeontological and geochemical analysis across a very complete boundary section. The sediments that make up the section were deposited in a shallow sea. Geochemical work has shown that iridium and other extra-terrestrially sourced elements were enriched at the boundary, so the position of the boundary and the evidence for the meteorite are confirmed.

The planktic foraminiferal assemblages[13] from the 2 metres of strata right below the event horizon show signs of what the authors called 'super stressed' conditions. These signs include:

---

[13] Foraminiferal species that float in the oceans.

- Intervals where carbonate – the mineral foraminifera use to build their shells – was dissolved;
- Poorly preserved (broken and fragmented) shells; and
- Dwarfed or misshapen shells, which are possibly a result of low oxygen levels.

This interval also recorded blooms of the foraminiferal genus *Guembelitria*. There were some samples examined where this genus makes up over 95% of the assemblage. *Guembelitria* is well known worldwide as a 'disaster species' – a genus that thrives when environmental conditions are stressful. It is one of the two polygons that bloom across the boundary on the Kerguelen Plateau (Figure 6.3). So, there is good palaeontological evidence for ecosystem stress prior to the impact.

The cause of the environmental stress is revealed in the geochemical data. It suggests that the carbon dioxide and sulphur dioxide released as the Deccan Traps were emplaced resulted in ocean acidification and periodic intervals of acid rain. The greenhouse effect that resulted from the addition of the volcanic gases into the atmosphere would lead to a warmer climate. In turn, the increased warmth would result in higher levels of weathering and runoff into the shallow sea, causing increases in turbidity. These conditions would result in the exclusion of most foraminiferal species, leaving *Guembelitria* to bloom. All in all, good evidence for stressful conditions prior to the extinction event. But Meghalaya is very close to the Deccan Traps, so perhaps it is an exception. We need to look further afield, say to Europe and ground zero for the meteorite model: the Gubbio section.

The Boundary in Gubbio

Because they contain minute magnetic particles, some sediments, as they accumulate, can record both the strength and direction of the Earth's magnetic field. Scientists retrieve these data and use them to help work out the age of a rock or to say something about the environment it was deposited in. One quantity they measure is the magnetic susceptibility, or the degree of magnetisation of a sediment. Alexandra Abrajevitch and a team of palaeomagicians[14] examined sediments from sections from Gubbio in Italy and Bidart in southern France. The standard explanation for this low-susceptibility interval is a two-step process. Firstly, water, after passing through large accumulations of dead organic matter that accumulated at the extinction layer, becomes strongly alkaline. Then, as this water percolates into the underlying sediments, it dissolves away the minerals that carried the magnetic signal, reducing the susceptibility.

Abrajevitch and her team's study of the Gubbio section confirmed the presence of the low-susceptibility layer underlying the boundary. However, when the researchers turned their attention to the boundary section from Bidart, they found something different. Yes, the low-susceptibility layer was there; however, just below the boundary but *above* the low-susceptibility layer, they found a brief interval where the magnetic signal returned to normal levels. This strongly implies that the lowered intensity layers aren't related to water passing through the organic remains left behind by the extinction. Instead, it suggests that something was going on before the impact. So far, this is interesting but not conclusive of anything. But if the change in magnetic susceptibility wasn't a result of alkaline water percolating through the overlying rocks, the palaeomagicians decided they would find out what it was due to.

Their study identified that the reduction in susceptibility was a result of an almost total loss of the biogenic magnetite. This is a mineral that many organisms (especially bacteria) carry around with them to help with finding their location or orientation. Abrajevitch's team suggested that the sharp reduction in biogenic magnetite indicates that there was a significant level of extinction prior to the impact that ended the Cretaceous. Again, this points the finger at longer-term ecological stress leading up to the main event.

---

[14] Jokingly, scientists who study the Earth's ancient magnetic field are often referred to as palaeomagicians. I'm not sure they like it.

Abrajevitch suggested that the stress was a result of an increase in ocean and atmospheric acidity caused by the Deccan Traps. This implies that the influence of the emplacement of the Traps extends as far as Europe. How far from the Deccan Traps can we go? How about Antarctica?

The Boundary in Antarctica

Sierra Petersen, Andrea Dutton, and Kyger Lohmann examined a complete Cretaceous–Paleogene boundary section from Seymour Island, which lies off the east coast of the Antarctic Peninsula. Using complex (but relatively reliable) geochemical proxies, they estimated the temperature of the area as it approached the end of the Cretaceous. They found two distinct increases in temperature. The oldest (a rise in temperature of about 8 degrees Celsius) coincides, as close as we can tell, with the beginning of the eruption of the Deccan Traps. Petersen, Dutton, and Lohmann went on to chemically analyse the shells of molluscs that were living in the latest part of the Cretaceous. Variations in the isotopes of carbon and oxygen within a single shell suggest that there was a reduction in seasonality, meaning that the difference between summer and winter temperatures was lessened. They also linked this pre-impact warming with an early pulse of extinctions. There is a second, smaller, temperature increase coincident with the meteorite impact and the final end-Cretaceous extinctions. But the important point is that even in Antarctica, there is likely to be a record of the emplacement of the Deccan Traps causing climate change, increasing environmental stress, and extinctions in the lead up to the impact event itself.

So where does all that leave us? I believe that the end of the Cretaceous was an accident waiting to happen. There is an initial, long-term build-up of environmental stress caused by the eruption of the Deccan Traps. In many ways, this is similar to the situation at the end of the Permian. However, because the Deccan Traps are smaller than the Siberian Traps, their effects may have been less severe. Nevertheless, the Earth's biota was struggling. Was the environmental forcing caused by the Deccan Traps enough to cause a mass extinction on its own? Could the ecosystem have recovered as the effects of the Deccan emplacement faded? We don't know, because at a critical point the impact of a meteorite delivered the *coup de grâce*, and most of life on Earth was extinguished. The big issue with the end-Cretaceous event isn't whether it was caused by either a volcano or an impact – they both happened. What we don't know is the relative contribution each made to the event.

Are all mass extinctions accidents waiting to happen? A build-up of longer-term ecological stress pushing life towards extinction but requiring a further trigger to deliver the final blow? For the end-Cretaceous event, this certainly seems to be the case. And we have seen that in the case of the end-Permian mass extinction, it seems likely that the environmental conditions that were largely a result of the formation of Pangaea predisposed the late Permian biota to mass extinction, but it took the emplacement of the Siberian Traps to push it over the edge into the extinction event.

Figure 6.1 demonstrates the link between many mass extinctions and the emplacement LIPs. It is still not certain whether an additional trigger like a meteorite impact or a deteriorating climate is needed to deliver the final blow.[15] Nevertheless, there is a lot of work being carried out, particularly with respect to the Big Five, so more clarity is almost certain to emerge.

Having spent quite a bit of time talking about what causes a mass extinction, it's time to move on past all the gloom and doom, and look at how life recovers afterwards.

---

[15] A recent report has implicated a supernova in causing the end-Devonian event alongside the emplacement of the Viluy Traps. See Fields and others (2020) in the Further Reading section.

7   Time Heals All: Recovering from a Mass Extinction

REPLACING SPECIES

When discussing a mass extinction, we generally focus on the species that went extinct and what triggered the event. This is what I've been concentrating on over the last two chapters. In this chapter, however, I'm going to look at what happens after a mass extinction and talk about how life on the planet recovers. What follows each of the massive events that defines a mass extinction is the gradual re-establishment of the Earth System's climatic equilibrium. In order for this to happen, the biosphere needs to recover from the damage created by the mass extinction, and this requires a rebuilding of the lost biodiversity. The recovery phase that follows a mass extinction takes a considerably longer time than the extinctions themselves: it takes place over many millions of years, as opposed to the thousands it took for some mass extinctions to kill off a sizable chunk of the Earth's biodiversity. In terms of evolution and the history of life on Earth, it could be argued that the recovery phase is at least as important as the loss of so many species.

Let's start with the obvious. No matter how catastrophic the effects of the various mass extinctions were, life did survive them all. The fact that you are reading this book is testament to the fact that all your ancestors have successfully made it through each and every mass extinction. The same is true, of course, for the ancestors of all 8.7 million species alive today. We may be living in life's margin of error, but we are good at it. Life's success in surviving catastrophic mass extinctions is illustrated in the estimates of the levels of ancient biodiversity I talked about in Chapter 4 (see Figures 4.3–4.6). They all clearly show that at no point over at least the last 600 million years of life's history did biodiversity drop to zero – a point where life and evolution would have to start again from scratch. We can also be fairly safe in assuming that the same applies to the whole 3.7 billion years that life has existed on Earth.

As well as showing us how successful all our ancestors were at surviving, the estimates of ancient biodiversity also demonstrate that mass extinctions are powerful engines of biological change. They have the ability to alter the composition of the planet's biodiversity on the largest scale. This is even reflected in some of the language used in some older textbooks (certainly the older ones that I used as an undergraduate) or in some articles aimed at the general public. The time period that followed the end-Cretaceous mass extinction, the Cenozoic Era, is commonly referred to as the 'Age of Mammals'. On the other hand, the Mesozoic Era which precedes the Cenozoic is commonly referred to as the 'Age of Reptiles'. These two labels reflect the change brought about by the end-Cretaceous mass extinction, from a terrestrial biota dominated by reptiles (dinosaurs) to one dominated by mammals. The estimates of ancient biodiversity quantitatively illustrate a parallel change – the biota that enters a mass extinction is not the same as the one that emerges. Mass extinctions accomplish this through wholesale emptying of environmental niches that had been occupied by the species that went extinct, and their refilling by a different set of species.

It is very difficult to predict which species will go extinct at a mass extinction. If the extinction event was solely a result of longer-term environmental forcing produced by the emplacement of a LIP such as the Deccan Traps, we might be able to make some sort of guess. In this case, we could document the abundances of various species as environmental conditions deteriorate before the mass extinction event itself. We could then speculate that those species that are undergoing a significant fall in abundance are the most likely to go extinct. We saw just this situation in the foraminiferal data from the Kerguelen Plateau (Figure 6.3), where the data show that most of the species that suffered a significant loss of abundance prior to the extinction event ultimately went on to disappear entirely from the fossil record. If, however, the cause of the mass extinction has no precursor events, which, for example, would be the case if the end-Cretaceous mass extinction was entirely the result of a meteorite impact, then it would be simply impossible to make any prediction of which species would be killed off.

After the mass extinction has removed what might be the majority of species on the planet, the recovery phase kicks in. Firstly, the rate of origination of new species increases significantly. Figure 5.2 showed John Alroy's estimate of the rate of extinction over time, with the Big Five mass extinctions standing out against a background of other lesser extinction events. I have reproduced it again here as Figure 7.1. This time, however, I have added to it his estimates of the rate at which new species evolve – the origination rate. The extinction curve is coloured grey, and the solid black line is the origination rate. Concentrating on the Big Five mass extinctions, you can see that the rate at which new species evolve peaks immediately following each of the Big Five mass extinctions. These new species are filling the empty environmental niches vacated by the species killed off during the extinction event. All the empty environmental niches represent a huge opportunity, and evolution quickly steps up a gear and fills them. It's not until evolution has managed to fill most of the available niches and the ecosystem starts to function 'normally' again that the rate of origination starts to fall back to 'normal' levels.

If it's difficult to pick the losers in a mass extinction, then it's impossible to pick the winners. I cannot see how we can predict in advance which species will flourish in the aftermath of a mass extinction. This is because the replacement of the huge numbers of extinct species killed off during a mass extinction is an opportunistic affair. Evolution will use whatever species survive the mass extinction to fill the vacant niches. This means that the species that evolve during the burst of origination that follows a mass extinction need not be related to the ones that went extinct. We could ask whether a time-travelling

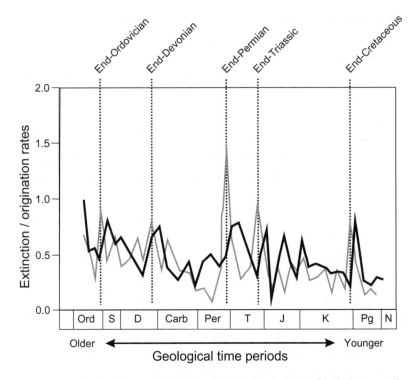

FIGURE 7.1 Extinction (grey line) and origination rates (black line) over the past 500 million years. The horizontal axis is geological time. The vertical axis is the rates of extinction and origination of genera per geological time interval. The Big Five mass extinctions are indicated by the dashed lines. Adapted with permission from Alroy (2008).

palaeontologist, sent back to the end of the Cretaceous, would be able to pick that the mammals scuttling around at the dinosaur's feet would be the ones to replace the reptiles in an upcoming mass extinction. I really doubt that they could, but the mammals did just that, eventually filling most of the ecological niches vacated by the reptiles.[1] During the refilling of empty niches that follows a mass extinction, evolution is able to produce a myriad of new and novel forms,

---

[1] Although just after the mass extinction the birds had a go at filling the vacant niches before the mammals finally won the day. In fact, birds are still doing pretty well (they are the 'dinosaurs' that made it).

restoring biodiversity and rebuilding the shattered biosphere. Indeed, some palaeontologists have suggested that most evolutionary novelties owe their existence to mass extinctions.

In the palaeontological literature, this repopulating of the planet and restoration of the biosphere is usually described as rapid, taking only a few million years. From the perspective of deep time, this *is* reasonably rapid, but on a human time scale it is a very long time indeed. There are two key reasons why the recovery takes so long. Firstly, it simply takes time for evolution to do its job: the production of new species is not a fast process. Secondly, the new and difficult environmental conditions such as greenhouse temperatures or acidic and anoxic oceans that triggered the mass extinction do not just go away immediately after the extinction events themselves. It takes time for the machinery of the Earth System to crank back into operation and restore the planet's climate to something close to its pre-mass-extinction state. Life is part of that restoration operation – and remember that following a mass extinction, life has just taken a huge hit.

ECOSYSTEM RECOVERY

Over the past 600 million years, the Earth System has successfully maintained planetary conditions that are suitable for the continuation of life. It is certainly true that climatic conditions have not stayed exactly the same for that whole period. For example, the planet has moved in and out of ice ages, gone through periods of exceptional warmth, and experienced intervals of heightened oxygen levels. But these have been transitory events – you can think of them as wobbles around a climatic equilibrium point, a pattern I showed in Figure 1.2. So, even though the Earth System isn't a perfect machine, in general it has managed to maintain conditions in a *relatively* constant state since the Cambrian; at least, a state in which life – despite ups and downs – has managed to continue.

Figure 1.2 also showed what happens when the machinery of the Earth System cannot maintain its balance. The application of significant environmental forcings to the Earth System can cause a sudden shift in the climatic equilibrium, which would be likely to generate a mass extinction. These forcings may be different for each mass extinction. In the case of the end-Permian event, it's the climatic damage caused by the emplacement of the Siberian Traps combined with the climatic stress caused by the presence of Pangaea. The end-Cretaceous extinctions are due to the environmental effects brought about by some combination of eruption of the Deccan Traps and a meteorite impact. The equilibrium shift caused by the environmental forcings results in the establishment of a new set of climatic conditions that much of the biosphere has not evolved to deal with. The end result is a massive die-off of species and the consequential collapse of the planet's ecosystem. Because we tend to concentrate on the extinctions, we tend to forget that the environmental forcings, and the disequilibrium conditions they cause, don't just go away as soon as the extinctions are over. Life has to deal with difficult or unusual environmental conditions for a considerable time following a mass extinction.

Before the Earth System can become fully operational again and regain its optimal climatic equilibrium, the biosphere, damaged by the mass extinction, needs to be repaired. This is because most of the important fluxes involved in maintaining the Earth System's equilibrium cycle through the biosphere. For the biosphere to support the flow of the fluxes through it, a high level of biodiversity and a fully functioning ecosystem are absolute necessities. However, following a mass extinction, the planet doesn't have a high level of biodiversity or a fully functioning ecosystem, which means that the fluxes are unable to operate properly. It is only when the forcings that initially triggered the mass extinction are removed that the planet's biodiversity is able to grow, allowing the global ecosystem to recover. As the ecosystem heals, the fluxes that are needed for the Earth System to operate will gradually start to come back on stream. As the recovery progresses, the machinery of the Earth System will begin to function 'properly', slowly undoing the damage caused by the mass extinction and eventually restoring a more benign climate. This phase of repair and reconstruction takes time, as one flux after another gradually starts to operate, altering the climatic conditions and optimising them for the continuance of the diversity of life on Earth.

In some well dated, well sampled, geological sections from across a mass extinction event, the fossil record actually documents the recovery of the ecosystem that follows a mass extinction. I want to look at two examples, both from perhaps the best studied mass extinction event: the end-Cretaceous.

In Aotearoa/New Zealand, Vivi Vajda and her colleagues have documented the changes in plant assemblages from across the Cretaceous–Paleogene boundary using fossil pollen and spores. The section they studied comes from the Moody Creek Mine on the west coast of the South Island of Aotearoa/New Zealand. It is well sampled and contains the peak in iridium that is associated with the meteorite impact, so we know exactly where to place the Cretaceous–Paleogene boundary. Prior to the extinction event, the area now occupied by the Moody Creek Mine supported a diverse plant population. It was dominated by conifer-like plants, but flowering plants and ferns were also abundant. Samples taken immediately following the extinction event show just how badly this flora was affected. They contain very little conifer pollen and no pollen of flowering plants. Instead, they are completely dominated by the spores of land ferns – in fact, land fern pollen makes up about 90% of the fossil assemblage. Today these sorts of ferns are seen as pioneer plants, the first to recolonise an area following some catastrophic event such as a fire. The situation in the earliest part of the Paleogene seems to be the same. The pre-extinction plant assemblages were devastated by the mass extinction, and the first plants to make it back were the pioneering ground ferns.

The return to the pre-extinction event conditions is reflected in the plant assemblages. Soon after the initial increase in ground ferns, they are supplanted by tree ferns. In turn, the tree ferns are replaced by the return of conifers. By this stage, the assemblage is starting to look a lot like the one that existed prior to the mass extinction. The flowering plants – which would complete the return to pre-extinction assemblages – don't arrive until much later. We don't know exactly how much later, since the section isn't well dated – but that's not the case with the next example.

In the Northern Hemisphere, Tyler Lyson and his colleagues analysed a geological section from Colorado. They documented the changes in both the plant community and the mammal population that followed the end-Cretaceous event. The section they used doesn't have the iridium peak to mark the boundary – but the fern spike, as identified in Aotearoa/New Zealand by Vajda, is present, allowing placement of the Cretaceous–Paleogene boundary. The section is also exceptionally well dated, so the authors can discuss the events that followed the mass extinction in terms of hundreds of thousands of years, not millions.

The authors describe the mammal assemblage that immediately followed the mass extinction as a 'disaster assemblage'. Its diversity is very low and consists entirely of small insect-eating species. This mammal disaster assemblage coincides precisely with the fern spike. Following the initial disaster assemblage, in a series of steps, both the size of the mammals and their level of diversity increases. At the same time, there are significant shifts within the plant community. These steps occur at about 100,000, 300,000, and 700,000 years following the mass extinction. Each of these recovery steps appears to be coincident with a brief rise in temperature, which the authors attribute to bursts of carbon dioxide from the ongoing eruption of the Deccan Traps.

Looking in a little detail at the structure of the recovery, at the 100,000-year step there is a marked increase in both the biodiversity and the size of the mammal species present. Within the 100,000 years following the mass extinction, mammals have almost recovered their pre-extinction size. Plants record a similar pattern of recovery to the one recorded by Vajda in Aotearoa/New Zealand, although the species involved are entirely different. The Colorado section records an increase in plant biodiversity following the fern-dominated ecosystem, reaching a maximum at the 300,000-year step.

While this is happening, the size of mammals increases. By the 300,000-year mark, some mammal species are three times the size they were prior to the extinction. Between 300,000 and 700,000 years, the size of mammals continues to increase. The fossils also indicate a changing feeding strategy. Species from the initial disaster assemblage were largely insect eaters. But over this interval more plant eaters appear, taking advantage of the increased diversity of vegetation available. Mammals reach their maximum size at 700,000 years. The appearance of the largest mammals at this step is thought to reflect the appearance of protein-rich legumes (peas, beans, clover, and the like), which occurs at the same time. In the Colorado section, the full recovery from the mass extinction is thought to have taken a further 300,000 years, so a million years in total.

EARTH SYSTEM SUCCESSION

Both of these examples show the slow rebuilding of a damaged ecosystem by the Earth System as it works to restore climatic equilibrium immediately following the extinction event. Both sections start with a record of low-diversity plant assemblages dominated by pioneering species such as ground ferns. As well as this low-diversity plant flora, the Colorado section also records a low-diversity mammal population composed of very small individuals. In the thousands or millions of years that follow the mass extinction, both sections show a steady increase in biodiversity and ecosystem complexity. The pattern of recovery documented in these two sections is not unique. Similar ecological successions following mass extinctions have been recorded in other geological sections from around the globe. All are startlingly similar to the ecological successions that follow modern (relatively short-term) disasters such as wildfires or floods. Recovery from modern disasters also starts with low diversity, often dwarfed assemblages of pioneer species that are followed by a slow rebuild as the complexity of the damaged ecosystem is restored. There are, of course, major differences. The rebuild that follows a mass extinction happens on a global scale, whereas a modern disaster is clearly a relatively local event. But the big difference is the time taken for the rebuild. A recovery from a modern disaster may take tens or hundreds of years, whereas the recovery from a mass extinction may take millions of years.

Nevertheless, the similarities between modern disaster recovery and what follows a mass extinction suggested to Pincelli Hull that, despite the massive difference in scale, the same basic ecological processes are taking place. She suggested that the rebuilding of the biosphere that follows a mass extinction represented a new type of ecological succession, one that parallels the ecological successions we see today following local disaster events, but operates over millions of years and on a global scale. To reflect this difference in scale, she suggested that the new succession after a mass extinction event be called an 'Earth System succession'.

As well as emphasising the importance of the interaction between the biosphere and the other 'spheres' of the Earth System, the recognition of this new ecological succession means that we shouldn't think of a mass extinction as being just about the species that go extinct. Each mass extinction is a two-part event that starts with a rapid burst of extinction, followed by a longer recovery phase. In terms of the history of life on Earth, I would certainly argue that the recovery phase is just as important as the loss of so many species.

Having said that, I want to devote the next chapter to a single very significant extinction event. Some palaeontologists don't consider it a full mass extinction, even though it does strongly affect the planet's biodiversity. It is an important event and worth spending time over. It is fairly recent, happening only 10,000 years or so ago, and humans are heavily implicated as being its cause.

8   The Late Quaternary Megafaunal Extinctions

In the mid-1970s, a lonely skeleton of a *Diprotodon* could be found on the ground floor of the South Australian museum. Solitary, it was tucked away in a corner near the main entrance and, as far as I could see, largely ignored by visitors – except by children who thought it was a dinosaur. It always looked a bit sad to me. As a young palaeontologist-in-training, my laboratory class spent an afternoon in the museum drawing its skull. I have reproduced my modest attempt in Figure 8.1.[1] Today, the skeleton has been moved to another, more prominent, display where it sits alongside some of its contemporaries.

Diprotodon was the largest marsupial that ever lived. There were several species (the exact number is still being debated), but they were all about the size of a modern rhinoceros. They were undoubtedly marsupials, with a pouch and wombat-like pigeon toes. There is nothing dainty about the skull, with its massive jaws and two huge incisor teeth extending forward from the lower jaw (its name, *Diprotodon*, means 'two first teeth'). Covered with fur, these giants grazed outback Australia during the late Pliocene and early

---

[1] As an aside, the point of the exercise was not to improve our drawing skills. We drew many skulls from a disparate group of animals. The aim was to make us aware that the same set of bones could be found in every skull we drew. They changed their shape and position, and sometimes several bones fused together – but we could recognise them every time. Features like these that can be recognised across a wide variety of animal groups are said to be 'homologous', and they indicate that all the groups shared a common evolutionary ancestor. The bones in your arm and in the wings of a bird are homologous, meaning that sometime in the very distant past birds and humans shared a single common ancestor.

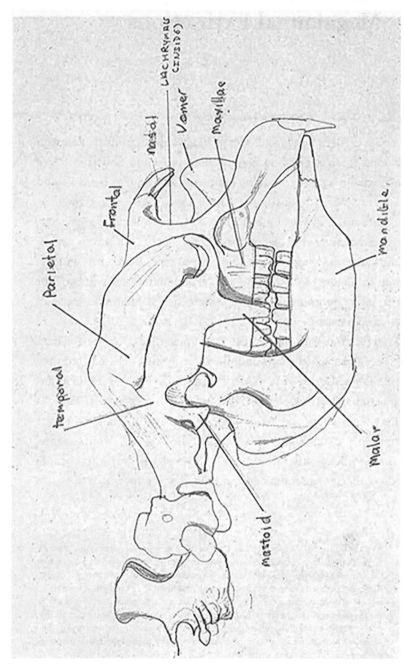

FIGURE 8.1 Drawing of a Diprotodon skull, by the author (when he was very young). Note the lack of a scale – the skull is in fact about 60 centimetres long.

Pleistocene. And they weren't alone. Alongside the diprotodons were other giants including huge kangaroos (*Procoptodon goliah*) that stood over 2 metres tall and weighed up to 200 kilograms. One of the largest carnivorous marsupials that has ever existed, *Thylacoleo carnifex*, was about 1.5 metres long and equipped with massive teeth for shearing flesh. A giant flightless bird, *Dromornis stirtoni*, was 3 metres tall and weighed up to 650 kilograms: some palaeontologists believe *Dromornis* was carnivorous. And then there was *Varanus priscus*, a gigantic carnivorous monitor lizard which grew to between 5 and 7 metres in length with an estimated weight of up to 2,000 kilograms.

At the same time as these giants were occupying Australia, in other parts of the world faunas dominated by large animals were also flourishing. In North America lived elephant-like mastodons, giant beavers, and dire wolves (George R. R. Martin didn't invent them for the *Game of Thrones*). Giant ground sloths browsed the trees of both North and South America. Herds of woolly mammoths wandered across Europe and Asia, alongside extinct species of rhinoceros and bison. In Aotearoa/New Zealand, the giant (and thankfully herbivorous) flightless moas were preyed upon by the huge Haast's eagle (Pouākai). Across the globe, there were many other species of giants, and taken together, this collection is referred to as the 'Quaternary megafauna'.[2] Most members of the megafauna are now extinct. On all the continents, except for Africa, most of the extinctions occurred sometime between about 50 thousand years ago and the beginning of the geological period called the Holocene, around 11 thousand years ago. Africa is the exception; there, the megafauna is largely intact. Today, in Africa, elephants, rhinos, hippopotamuses, and the big cats are the remnants of an only slightly more diverse Quaternary African megafauna. Away from the continents, megafaunal extinctions are

---

[2] An animal is classified as being part of the megafauna if, in life, it weighed over 44 kilograms. This seems like an odd number: why 44, not 45 or 50? I suspect (but can't prove) it's because whoever first defined the term megafauna was using imperial measurements – 44 kilograms is close to 100 pounds.

Table 8.1 *Percentage of mammal genera going extinct during the megafaunal extinctions*

Data from Koch and Barnosky (2006). Those authors note that in Australia seven genera of megafaunal reptiles and birds also went extinct.

Continents	Number of megafaunal genera	Number of extinct megafaunal genera	Number of Holocene megafaunal survivors	Percentage of extinct megafaunal genera
Africa	48	10	38	21
Australia	16	14	2	88
Eurasia	26	9	17	35
North America	47	34	13	72
South America	60	50	10	83

also recorded on islands such as Madagascar and Aotearoa/New Zealand. On these islands, the intensity of the extinctions was very high, but they occurred much later than those on the continents, well after 10 thousand years ago. There is a reason for this, and I'll go into that shortly. But before I do, we need to decide whether the megafaunal extinctions do represent a *mass* extinction.

The first thing we need to look at is the scale of the extinction event. Table 8.1 gives some idea of the numbers involved with the continental megafaunal extinctions. It shows the number of genera going extinct on each continent. As I have noted before, this means that the number of species going extinct will be much higher. This was clearly a significant event. However, although the extinctions were global in nature, the data in Table 8.1 show that the intensity of the extinctions varied from continent to continent.

As I noted above, Africa suffered the least: of the forty-eight megafaunal genera that lived during the Pleistocene, ten (21%) died out during the extinction event, and thirty-eight survived into the

Holocene. Over the same time interval, Australia fared the worst, with an 88% rate; only two genera from the original sixteen managed to survive into the Holocene. The Australian extinctions were actually worse than the table shows, as it lists only genera of mammals that went extinct. Australia also lost seven additional megafaunal genera, representing reptiles and birds. South America wasn't far behind Australia (in fact, some reports put it ahead). It starts with the highest diversity of megafaunal genera of all the continents, but only ten survive today, indicating a very high extinction rate of 83%.

So, clearly a large, global extinction event – but can we call it a mass extinction? In Chapter 6, I provided a definition that may help us decide. Essentially, in order to gain the epithet of a mass extinction, any extinction event must satisfy two criteria. The first is that the event must take place over a very short interval of deep time. As we will see, the dating of the megafaunal extinctions is problematic; however, we can be certain that they took place over a period of about 40 thousand years. This makes the event geologically rapid, and that satisfies the first criterion. The second criterion is that it had to represent the loss of more than one geographically widespread higher taxon, resulting in a drop in overall biodiversity. The data indicate that the megafauna extinction did involve more than one genus, and it did affect animal populations from across the globe, resulting in a significant lowering of biodiversity, fulfilling the second criterion. Taken with the rapidity of the extinction event, my broad definition of a mass extinction is satisfied. But is it enough?

The notion that the megafaunal extinctions represent the youngest mass extinction in the fossil record has been rejected by a number of scientists. They note that the ancient mass extinctions such as the Big Five cut a swath of destruction across the planet's entire biota. Reptiles, mammals, plants, and insects were just some of the groups that were decimated during the extinction events. In the case of the megafaunal extinctions, only a single animal group was affected: all the genera that went extinct are vertebrates. The limited impact of the extinctions certainly sets it apart from the other ancient mass extinctions.

So, despite the megafaunal extinctions meeting my definition of a mass extinction, the fact that it affects such a limited range of animal species makes some palaeontologists uncomfortable in calling it a mass extinction. In some ways, it's a bit like the situation I discussed in Chapter 1. In that case, I rejected the term mass extinction for the current Anthropocene extinctions in favour of defaunation. This wasn't because I didn't believe that we are at least at the beginning of a mass extinction. I chose to use the term defaunation because, firstly, the level of actual extinctions is considerably lower than any of the ancient mass extinctions; and secondly, the concept of defaunation includes both the species that are extinct and those that are under threat of extinction, providing us with a more complete picture of the situation we find ourselves in. But that logic doesn't apply to the Quaternary megafauna. So, while I recognise that it's small (relatively speaking) and affects only a single taxonomic group, I still think it is best described as a mass extinction, making it the youngest on record. While the megafaunal mass extinction pales in comparison to the other ancient mass extinctions, it remains a very significant event – and we humans, *Homo sapiens*, lived through the whole thing.

As *Homo sapiens* emerged from Africa and spread through the Levant, Europe, and Asia, we encountered, and interbred with, other human species whose ancestors had left Africa before us. As a result, many of us carry with us the genes of *Homo neanderthalensis*[3] and a group referred to as the Denisovans. It is also possible that we interbred with several other so-called ghost species. These are species of humans for whose existence we have DNA evidence – but no fossils. As the geographical range of our species expanded, in addition to

---

[3] I am writing this during the COVID-19 pandemic, and some reports suggest that the presence of certain Neanderthal genes in some people's genomes may place them at increased risk of the disease. See Zeberg and Pääbo (2020) in the Further Reading section.

meeting our sister species we also encountered the megafauna. Because we were present as the megafauna went extinct, there are suggestions that we may be either wholly, or at least in part, the cause of the extinctions. The implication that we are involved in some way with this mass extinction is the reason I'm devoting so much time discussing it. In particular, I'm going to concentrate on the continental megafaunal extinctions listed in Table 8.1. However, before I do that, I am going to take a small digression and discuss the megafaunal extinctions on small islands.

SMALL ISLAND EXTINCTIONS

As we will see, intense debate rages about what caused the continental megafaunal extinctions. In the case of small, isolated islands, however, there is no doubt about what drove the megafaunal extinctions: they are a direct result of human occupation. The reason that these extinctions occurred so long after the continental extinction lies in the relative isolation of small islands and the length of time it took humans to get there.

When humans migrate to an island and establish a settlement, as well as clearing land for settlement and agriculture they bring with them foreign pests such as rats and mice. In addition, the need to feed an expanding population can lead to overhunting. Human occupation could trigger an extinction event on its own. However, there are also aspects of island ecology that makes the indigenous fauna more vulnerable to human-induced extinctions. Firstly, island faunas usually include a high number of endemic species. An endemic species is simply one that is restricted to a certain area, in this case an island. In continental areas where non-endemic species are more common, the level of biodiversity can be more easily maintained. This is because when a species goes extinct in one area it can be replaced by migration from another. However, in the case of an island with lots of endemic species, because they are found nowhere else, when one goes extinct it can't be replaced, and island diversity falls.

Secondly, prior to the arrival of humans, most small islands lacked any significant predators.[4] As a result, indigenous species would not be used to being preyed on and would be less likely to take evasive action when faced with newly arrived and well equipped hunters. It is said that the local fauna would be *naïve*. This is an important concept, one we will come back to when discussing the continental extinctions. An example of this is the modern dodo on Mauritius. Being unafraid of the marauding sailors made them particularly easy to kill off.[5] Overhunting, land clearance, and invasive species coupled with endemic, naïve species could quickly result in extremely rapid extinctions. In Aotearoa/New Zealand, for example, all the moa species were wiped out in the 150 to 200 years following human settlement from Polynesia.

The strong linkage between human settlement and island extinctions helps to explain their lateness. Remember, they usually occurred thousands of years after the continental extinctions were over. This is because the date of the extinctions on small islands simply reflects the date of human settlement. For example, if we look at the peopling of the remote islands of the Pacific, Polynesian occupation couldn't take place until after they had developed seagoing waka (canoes) and the amazing navigation skills they needed to make the migration. As a result, human occupation and island extinctions occurred well after those on the continents. In other words, an island's isolation simply delayed the arrival of *Homo sapiens* and the almost inevitable wave of extinctions that accompanied them.

CONTINENTAL EXTINCTIONS – A CONFUSION OF CAUSES

Moving onto the continental megafaunal extinctions, you would think that because this is the youngest mass extinction and very close to the present day, it would be easy to work out exactly what caused

---

[4] The giant Haast's eagle (Pouākai) in Aotearoa/New Zealand with its 3-metre wingspan is an obvious exception.

[5] Although this almost certainly will not be the whole story: land clearance and the introduction of cats, dogs, and pigs to Mauritius also contributed.

it. Frustratingly, that is not the case, and palaeontologists are still actively debating the issue. The majority of suggestions for what killed the megafauna fall into two distinct camps: either the megafaunal extinction was due entirely to human activity, or it was due entirely to a rapidly changing climate. The scientific literature addressing this debate is extremely polarised, with supporters of each explanation producing papers that declare, with absolute certainty, that their perspective is the only option. Although the debate doesn't seem to be as divisive as the one that is going on about the meteorite at the end of Cretaceous, both sides are undoubtedly talking past each other. In particular, the group blaming humans for the megafaunal extinctions seem to be particularly negative about accepting any alternative viewpoint. The impression I have is that they believe that if humans are let off the hook here, then our part in the Anthropocene defaunation can be played down. This isn't a viewpoint I share.

As a counterpoint to these two somewhat entrenched positions, there is a third group of palaeontologists and geologists who take the middle ground. This group believes that explaining the megafaunal extinctions isn't a choice between either human activity or climate change. The extinctions, they say, are a result of a combination of the two. I said earlier in the book that I am an inveterate fence-sitter, so it should not surprise anyone to hear that I have a lot of sympathy for this approach.

There is another, extremely controversial, model that has been suggested to explain the megafaunal mass extinction – that the impact of a meteorite (or comet) played a major role in the megafaunal mass extinction. We have already seen that an extra-terrestrial impact was heavily involved in the end-Cretaceous mass extinction 65 million years ago, so it's not beyond our imagination to see a role for an impact in the megafaunal mass extinction. Defenders of this model are in a somewhat beleaguered minority – but support for the model is steadfast, so I need to discuss it.

That leaves us with four models to explain the megafaunal extinctions: cosmic impact, humans on their own, changing climate on its own, or a combination of climate and humans. All four rely heavily on accurate dating of the climatic events, human arrival, and the extinctions. The dates of key climatic events, which on the surface would appear to be difficult to date, are actually very well known. This is because we have excellent climatic records from well dated, complete ice cores for this time period. Unfortunately, accurate dating of human arrival and the megafaunal extinctions is just what we do not have.

DATING PROBLEMS

One problem we face when it comes to dating human and megafaunal fossils using traditional geological techniques is the Signor–Lipps effect. We have encountered this effect before, in the discussion of the end-Cretaceous event (see Figure 6.4), where I showed how a sharp mass extinction could be artificially smeared out over time, making it look gradual. This is a result of the lack of certainty that we encounter when trying to date the exact top or base of a species' range. In the case of the end-Cretaceous mass extinction, I argued that the effect was minimised, because in the example I showed the samples were very closely spaced, and the commonly used micro-fossils had a fantastic fossil record (see Figure 6.3). However, for the megafaunal extinctions, we are dealing with a situation where very patchy sampling is coupled with two fossil groups (hominins and the megafauna) that have a very incomplete fossil record. This is the perfect setting for a strong Signor–Lipps effect. It means that we don't know whether we are dealing with the true first appearance or true last appearance of a fossil, and therefore we cannot be certain of the dates for those events. In the case of the dating of human arrival, we do have new techniques available. The emerging science of the analysis of ancient genetic data can provide arrival dates.[6]

---

[6] If you are interested in the power of genetic analysis in aiding our understanding of human evolution, I recommend *Neanderthal Man* by Savante Pääbo and *Who We Are and How We Got Here* by David Reich – both astonishing books.

While this avoids the Signor–Lipps effect, it does mean that we are open to the vagaries of interpretation of genetic analyses.

In Table 8.2, I've listed some of the available dates for both the arrival of humans and the timing of the extinctions. These dates should be considered to be very, very approximate and treated with caution. An examination of the literature on the megafaunal extinctions yields a multitude of often conflicting dates for the both the extinctions and arrival of *Homo sapiens.* This makes any analysis of the causes of the megafaunal mass extinction extremely difficult. I have used the available genetic dates for the dates of human arrival. The dates used for the extinctions come from a paper by Koch and Barnosky published in 2006. This makes them relatively old data – but it does mean that I am taking the dates from a single, reliable source.

Modern humans first settled permanently in Eurasia between about 50 thousand and 60 thousand years ago. However, there are a few locations where fossils of archaic forms of *Homo sapiens* have been found that date back to over 200 thousand years. Aboriginal Australians reached northern Australia about 55 thousand years ago, but it took an additional 15 thousand years or so to occupy the southern part of the continent. The early American Clovis people had entered North America just after 13 thousand years ago. There is, however, increasing evidence for an earlier, pre-Clovis occupation. These people seem to have arrived around 2 thousand years before the establishment of the Clovis culture. The date of human arrival into South America is much more uncertain. Genetic data suggest that people had reached that continent about 14.6 thousand years ago, pre-dating the Clovis people. To muddy the waters further, a recent publication by a group led by Ciprian Ardelean reported evidence of human occupation in the Chiquihuite Cave in Mexico that pushes human occupation of the area back beyond 30 thousand years.[7] Needless to say, this is a controversial result and will require corroboration. Until that arrives, I'll stick with the date in Table 8.2.

---

[7] See Ardelean and others (2020) in the Further Reading section.

Table 8.2 *Timing of megafaunal extinctions and human arrivals*

All dates (in years ago) should be considered very approximate. Data from: Koch and Barnosky (2006), O'Connell and others (2018), Pääbo (2014), Reich (2018), Saltré and others (2016), Tobler and others (2017).

Continent	Arrival of *Homo sapiens*	Extinctions start	Extinctions end	Comments
Eurasia	By >50,000 years*	17,000 years	10,000 years	Some mammoth species survive to about 4,000 years – possibly younger.
Australia	By 50,000 years	72,000 years	40,000 years	Humans may not have reached southern Australia until between 49,000 and 45,000 years.
North America	<13,000 years	13,500 years	11,500 years	Increasing evidence of an older arrival of humans
South America	By 14,600 years	?18,000 years	11,500 years	Dating very uncertain. Some controversial dates extend human occupation back significantly further

*This represents the 'main wave' of *H. sapiens* into Eurasia. There are fossils available suggesting that archaic forms of our species may have been in the area before about 200 ka.

The relationship between the timing of the extinctions and human arrival differs from continent to continent. In Eurasia, the extinctions peaked in intensity between 15 and 10 thousand years ago, although there was an earlier pulse between 50 and 24 thousand years ago. Mammoths survived until relatively recently. Mammoth fossils dated at between 4 thousand and 1.6 thousand years old have been recovered from Arctic islands north of the Russian Republic, suggesting that a small population survived at high latitudes. In Australia, the extinctions started well before the arrival of humans but reached a peak around 40 thousand years ago. In North America, the extinctions appear to start at about the same time as the Clovis people arrive and are essentially over within 2 thousand years. The extinctions in South America start around 3 thousand years after humans arrive and are essentially over around 3 thousand later.

Despite all the uncertainties and contradictions in the literature around the relative timing of the megafaunal extinctions and the arrival of humans, these dates are the cornerstones of all our efforts to explain the event. However, despite the limitations of dates, in the remainder of this chapter I want to assess each of the four models mentioned above.

A METEORITE (OR COMET) DID IT

The initial impact scenario suggested that, about 12 thousand years ago, a meteorite hit the planet in the Great Lakes region of North America, which at the time was covered by the immense Laurentide ice sheet. Although the proposed meteorite was a good deal smaller than the one at the end of the Cretaceous, the results were thought to be similar. Debris from the impact thrown into the atmosphere reduced the level of sunlight, which caused a sharp drop in global temperatures, triggering the extinctions. In other words, the impact model suggests that climate change did kill off the megafauna – but the trigger for the climatic change was an extra-terrestrial impact. We have known for a long time that there is a sharp climatic cooling around 12 thousand years ago. It was first recognised in the early 1900s, based on studies of fossil pollen. It's known as the Younger Dryas[8] event. Since its initial recognition, it has been recorded in many sites from the Northern Hemisphere.

The lack of a crater near the proposed impact site in the Great Lakes region was a problem. To get around this, it was suggested that instead of a single impact, a large comet or meteor disintegrated as it entered the Earth's atmosphere, with fragments striking North and South America, Europe, and Asia. There is even a report that a Palaeolithic village in Syria was destroyed when it was hit by one of the extra-terrestrial fragments. However, in 2018 Kurt Kjær and his colleagues announced the discovery of a possible impact site.

Kjær's team located a crater, some 31 kilometres in diameter, beneath a glacier in Greenland. However, to confirm it as impact site for the meteorite that triggered the Younger Dryas cooling, the crater must have been formed at the start of the event, around 12 thousand years ago. As we have seen, calculating the age of a crater is difficult – and in this case it was made even harder because the crater was sandwiched between rocks that the meteorite hit, which can be dir-ectly dated, and the glacial ice, which is difficult to date. As a result, the authors were circumspect when calculating an age for the crater. They noted that the only thing they could be certain of was that the crater was younger than the rocks it impacted (they are billions of years old) and older than the age of the ice that covers it. They did note that there was a higher heat flow from the crater than the surrounding area, suggesting that it was still cooling from a relatively recent impact, but were unable to estimate exactly how recent.

Based on the evidence they had – the age of the covering ice based on modelling of its flow and an analysis of the crater's morphology – they suggested that the impact occurred during the Pleistocene. This means that it happened somewhere between 2.5 million years and

---

[8] It is named after an alpine flower, *Dryas octopetala*. Its pollen is common in deposits formed during glacial periods. And yes, there is an Older Dryas cooling.

10 thousand years ago. However, after this careful assessment of the crater's age, they then noted that an impact event of the size needed to produce the crater in Greenland would have resulted in a significant change in global climate. This appeared to be a subtle hint linking the impact with the Younger Dryas. Suddenly there were stories everywhere, from the scientific literature to the popular press, that the smoking gun had been found, and the megafaunal mass extinction and the Younger Dryas cooling were the result of an extra-terrestrial impact. However, until the crater's age is properly tied down, the link between it and the Younger Dryas cooling must be considered speculative. Best just to stick with the available evidence that the crater was formed sometime during the 2 million years or so of the Pleistocene. This broad age is certainly not precise enough to definitively link the impact to the Younger Dryas climate event.

However, proponents of the impact-induced Younger Dryas model don't just rely on a possible crater to support their argument. They present a wide range of evidence from sediments dated to the beginning of the Younger Dryas across a number of sites. This evidence includes:

- Abundant nanodiamonds – tiny diamonds thought to be formed by the shock of the impact.
- Abundant tiny spheres of varying composition that are again thought to be created during impact.
- A spike in the level of platinum. Like iridium, an increase in the levels of this element is common at impact sites.
- Evidence of massive wildfires.

However, for every piece of evidence offered in support of the impact model, there is also evidence disputing it. In 2011, a group headed by Nicholas Pinter published a paper severely critical of the impact-driven extinctions.[9] Pinter and his co-authors questioned the identification of

[9] They gave their paper the catchy title *The Younger Dryas impact hypothesis: A requiem.*

the nanodiamonds and suggested a biological origin for at least some of the spheres. The authors also noted that some records of elevated levels of platinum couldn't be reproduced in later studies. They went on to suggest that there may be alternative sources of platinum other than an impact. The evidence presented in support of post-impact wildfires is dismissed as insubstantial. Pretty damning stuff.

One of the problems I have with the impact model, not mentioned by Pinter, is the simple fact that the Younger Dryas cooling is not recorded in sites from the Southern Hemisphere. In fact, as the Northern Hemisphere is cooling during the Younger Dryas, the Southern Hemisphere is warming (I'll talk more about this below). Can an impact produce completely different climatic effects on either side of the planet at the same time? If the cooling is a result of debris being kicked into the atmosphere, I would have thought that the effect would be a global-wide cooling. To counter this argument, supporters of the impact model suggest that the melting of the Laurentide ice sheet caused by the impact released huge amounts of fresh water into the world's oceans, which changed global circulation, preventing the cooling in the Southern Hemisphere. This idea remains untested.

Because the evidence for the impact is still widely questioned, I'm going to put it to one side. In the future I may come to regret this; the Younger Dryas impact model has vociferous champions, and additional work may provide conclusive evidence. In that case, I will be happy to re-examine my position.

HUMANS DID IT

The initial model for a human-driven megafaunal extinction was developed in the 1960s by Paul Martin and his colleagues. It was originally referred to as the Pleistocene overkill model. However, it quickly gained a new name – Blitzkrieg.[10] Martin's Blitzkrieg model

---

[10] I'm not sure that Martin ever called it Blitzkrieg – but he did write a paper called *40,000 years of extinction on the 'planet of doom'*, which is melodramatic enough.

was essentially the small island extinction model (extremely rapid extinctions immediately following human arrival) that we discussed earlier, but on a continental scale.

Under Blitzkrieg, a wave of *Homo sapiens* – intelligent hunters, fully equipped with the latest in Palaeolithic hunting technology – swept out of Africa and rapidly spread from continent to continent. As each new continent was occupied, they encountered the indigenous megafauna, which, because it had never encountered these smart new hunters before, was 'naïve' and thus easy to kill. The result was overkill, and the humans quickly drove the megafauna to extinction. Blitzkrieg explained quite nicely why the ages of the extinctions were different in each continent; as in the case of island extinctions, they simply reflected the timing of human migration. It also explained why Africa wasn't so badly affected. Here, the megafauna and humans had evolved alongside each other. The African megafauna wasn't naïve – it had learned how to live alongside and avoid the wily hunters.

The Blitzkrieg model has some significant problems, one of them being the question of the megafauna's naivety. It is assumed that when the hunters arrived in a continent the megafauna was unafraid of the new species moving in and essentially waited, dodo-like, to be killed off. That may be the case for a small island where we are dealing with a highly endemic fauna with no pre-existing large predators. It's difficult to see how it would work on a continent where the level of endemism was low and significant predators were already present. One would have thought that this would make the fauna more difficult to drive to complete extinction by human over-predation, since the fauna would have developed some behaviours to avoid being eaten well before humans got there. Studies of modern animals confirm this: they suggest that if a species is used to some predators but naïve when confronted with a newly arrived threat, it doesn't stay that way for long. Within a generation or two, the formerly naïve species would have developed the skills needed to help it avoid predation.

However, the biggest problem is timing. Blitzkrieg predicts that (1) the extinctions should start very soon after human arrival and (2) the actual time taken for the fauna to go extinct should be very short. Based on the dates provided in Table 8.2 (but remember I have already expressed some reservations about the data), I'm not convinced that they offer any real support for the model. The only likely candidates are North and South America – and that's a bit of a stretch. In Australia and Eurasia, humans and megafauna co-existed for thousands of years, and the extinctions were relatively slow.

To explain the time lag between human arrival and the extinctions, humans-only proponents suggest that it may simply reflect the fact that it takes time for a human population to establish itself and make its presence felt across an entire continent: in effect, a slow Blitzkrieg. The length of time for the extinctions to take place, as opposed to the near-instantaneous event required by the fast Blitzkrieg model, is explained by two factors: (1) the fauna itself would have had fewer endemic species, making it more difficult (and therefore take longer) to drive it to extinction; and (2) the first humans to meet the megafauna had not developed an agriculture-based lifestyle. This meant that they wouldn't have to embark on a programme of land clearance to provide space to grow their crops.

The difficulty with the timing creates enough uncertainty to make me more than a little suspicious of the idea that humans alone could have caused the extinctions. In addition, we know that shortly after the main exodus of humans out of Africa about 60 thousand years ago, and while the megafauna was going extinct, the planet's climate entered a highly dynamic phase. I don't think we can ignore the possibility that a dynamic climate contributed to the megafaunal extinctions.

CLIMATE DID IT

Could climate alone be responsible for the extinctions? It has certainly been very dynamic over the past 20 thousand years. But this

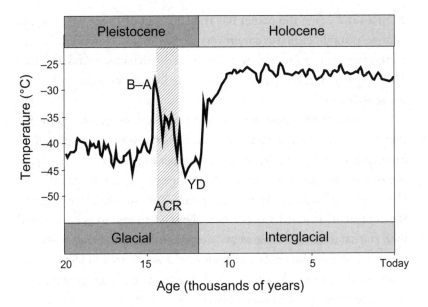

FIGURE 8.2 Northern Hemisphere temperature record of the last 20 thousand years. Data are from the Greenland GISP-2 ice core. B–A is the Bølling–Allerød event, YD is the Younger Dryas event. The shaded area represents the Antarctic Cold Reversal (ACR); note that this is a Southern Hemisphere event. Adapted from Platt, Haber, and others, Wikimedia CC-BY-SA 4.0.

short interval is only part of a much longer story of climate change – which we need to touch on only briefly.

Sometime around 2.6 million years ago, the Earth started to cycle through a series of cold glacials (ice ages) and warmer interglacials. Many times, over the last 2.6 million years, great ice sheets grew across continents, then melted away. As a student, I learned that there were only four great glacial periods in Europe,[11] but we now know there were many more. In fact, ice sheets have come and gone in a regular pattern that is closely related to the Milanković orbital cycles I mentioned in Chapter 5. Over the past

---

[11] And we had to learn the name of each of them. Living in Australia, that seemed to be a bit pointless and I promptly forgot them.

20 thousand years, the planet has transited out of an ice age into the relatively stable climatic conditions we see today. However, the climate over that transition was particularly dynamic. And it's this dynamism that some palaeontologists believe is enough to kill off the megafauna.

Concentrating now on the past 20 thousand years, Figure 8.2 shows a record of temperature over the transition from glacial to interglacial conditions. Because the temperature curve in the figure is based on data from an ice core from Greenland, it reflects climatic changes in the Northern Hemisphere only. The amount of ice covering the continents reached a maximum about 21 thousand years ago and then started to decrease. However, despite the retreat of the ice sheets, Figure 8.2 shows that cold conditions continued to around 15 thousand years. At that point, climate suddenly warmed, reaching temperatures similar to those we see today: this is called the Bølling–Allerød event. But the warmth didn't last. At about 12 thousand years ago, the climate cooled sharply again. This is the Younger Dryas event that some believe was triggered by an extra-terrestrial impact. Following the Younger Dryas, temperatures in the Northern Hemisphere rose again, and the planet entered the current interglacial, where climate has been stable for the past 10 thousand years. Compared with previous interglacials, this is a long period of stability.

Things were different in the Southern Hemisphere. Analysis of ice cores from Antarctica do not show any evidence of the Bølling–Allerød warming or Younger Dryas cooling events. Instead, the situation is almost exactly reversed – the climate warmed steadily following the glacial maximum except for an interval called the Antarctic Cold Reversal (marked by the shaded box labelled ACR in Figure 8.2) when cold suddenly returns. This event starts while the Northern Hemisphere is warming during the Bølling–Allerød and ends while the Northern Hemisphere is the grip of the cold Younger Dryas. Following the Antarctic Cold Reversal, data from the Southern Hemisphere record the same long stable interglacial as the Northern Hemisphere.

The end of the last glacial period and the beginning of the current interglacial occurs at about 11 thousand years ago. This event is used to mark the beginning of the interval of geological time we call the Holocene. As noted above, for the last 11 thousand years, climate has been remarkably stable – that is, right up until the impact of human activity that defines the Anthropocene. Except for Australia, where the megafauna was extinct well before the last glacial maximum, and the odd mammoth found in the high northern latitudes of Eurasia, most of the megafaunal extinctions seem to be associated with this dynamic climatic transition. This suggests to supporters of the climate-only model (and those that believe a meteorite was involved) that the megafauna simply couldn't survive such rapid shifts in climate. The Australian megafaunal extinctions that ended well before the last glacial maximum are put down to increasingly arid conditions on the continent about 40 thousand years ago.

Setting aside the issue of the Australian extinction dates, the greatest problem I see with the climate-only model is why the megafauna didn't go extinct earlier. There have been many other interglacials as the planet passed through numerous glacial and interglacial cycles. Indeed, some interglacials have been warmer than the one we are living in today, although none has lasted as long. Yet the megafauna lived through glacial cycle after glacial cycle, through intervals of rapid warming and cooling, with no mass extinction. Why should the megafauna wait until the last 11 thousand years or so to go extinct? Looking at the glacial cycles over the last 2.6 million years, there appears to be nothing special climatically about the lead up to the Holocene. There is of course one significant difference between early climatic fluctuations and the one that happened over the last 20 thousand years, and that is the coincident presence of increasing numbers of humans on the planet.

As was the case with the end-Cretaceous mass extinction event where I argued that to ignore the coincidence of the emplacement of the Deccan Traps and the meteorite impact was an error, I think the same logic applies in the case of the megafauna. We have two major events – one biological (the appearance of *Homo sapiens*), the other climatic (a glacial/interglacial transition) – occurring at the same time as a mass extinction. To me, this strongly implies that both are involved. However, to make a convincing case for a combination of humans and climate as twin drivers of the megafaunal extinction, we need more than simply their synchronicity; we need to show that both are required to trigger the mass extinction.

HUMANS PLUS CLIMATE EQUALS EXTINCTION?

There have been a number of suggestions about how the combination of climate and human activity could come together to trigger the megafaunal extinctions. One of the most elegant is recent work by a group led by Daniel Mann in 2019. They came up with a novel approach to link climate and human arrival in a coherent single explanation. As we will see, it has its problems, but it's worth serious consideration.

Surprisingly, Mann and his co-authors suggest that the climatic changes that occurred immediately prior to the Holocene, such as the Younger Dryas and the Antarctic Cold Reversal, weren't directly involved in the extinctions. Instead, they attribute the underlying cause of the extinctions to the change towards long-term climate *stability* at the beginning of the Holocene. The reason for the seemingly strange notion that climatic stability contributed to the extinctions is the recognition that the megafauna evolved at the same time that the Earth was cycling through glacial and interglacial periods. Natural selection had prepared them to cope with the ecological *instability brough about by the extensive glaciations*: but they weren't prepared for *stability brought about by the somewhat briefer interglacials.*

As we have seen, prior to the beginning of the Holocene, the planet cycled through glacial and interglacial intervals at a millennial-scale time frame. The extensive glacial intervals created a constantly changing patchwork of terrestrial environments. During warmer, more stable interglacials, this patchwork of environments gave way to what we would consider more normal, stable environmental conditions. Because the glacial periods were more extensive, the megafauna evolved to cope with this patchwork environment. However, during the interglacial periods, the megafauna weren't equipped to deal with the stable 'normal' environment. Mann suggested that this would result in a decline in numbers and perhaps severe threat of extinction – but they did not go extinct. Instead, they managed to pull through and survive. Once the interglacial was over, glacial conditions returned, the patchwork environmental conditions resumed, and the megafauna recovered.

Mann and his group noted that although the current interglacial is the longest on record and would have put the megafauna under severe stress, the megafauna didn't die out gradually as the stable climate dragged on. Instead, most of the extinctions are clustered near the beginning of the current interglacial. This raises the question: what made the beginning of this interglacial different to all the previous ones? Again, the obvious answer is the presence of humans. This suggests that it was the spread of humans combined with a megafauna that was stressed because of the stable environmental conditions that triggered the mass extinction. That would make it a two-part mass extinction, like the end-Permian and end- Cretaceous events. I think that this is a very appealing idea, although as Mann and his team acknowledged, there are still timing problems. In Australia, the extinctions were over by 32 thousand years ago, well before the Holocene, and in Africa, the megafauna largely still survives.

In the case of Australia, Mann and his co-authors suggest that the drying up of the continent that started about 40 thousand years ago provided the stability needed to set the megafauna up for extinction. Humans were also already present, having arrived about 55 thousand years ago. As with the later continental extinctions, this combination of a fauna under stress and the presence of humans triggered the extinctions.

In Africa, they argue that there may be at least two reasons why the megafauna survived there. Firstly, they invoke the same explanation as the Blitzkrieg model: the African megafauna wasn't naïve.

Unlike the rest of the world where humans are added to the ecosystem, in Africa the megafauna had evolved alongside humans and developed strategies to survive the increasing sophistication of human hunters. The second suggestion is that the African climate never achieved the climatic stability during the Holocene that was enjoyed by the rest of the world. In other words, that shifting patchwork of environments that the megafauna evolved to cope with persisted in Africa, allowing its survival. Mann and co-workers attribute the continuation of the unstable patchwork of environments to the continent's arid mid-latitude ecosystems being susceptible to boom and bust cycles of productivity.

I do believe that the notion that the megafaunal extinctions were due to either climate or human activities to be a false dichotomy: given the data involved, surely both must play a role. Mann may not have all the answers, but it is a lovely attempt at bringing all the pieces of the puzzle together.

The megafaunal mass extinction and the current Anthropocene defaunation are two events separated by around 11 thousand years: in terms of deep time, just a heartbeat. In both cases, humans are involved. Perhaps we should consider the megafaunal extinction as simply the beginning of the Anthropocene defaunation? As noted above, researchers who believe that humans are wholly responsible for the megafaunal extinctions seem to be particularly influenced by this notion. However, we almost certainly played a supporting role in the megafaunal mass extinction alongside a changing climate. In the case of the Anthropocene defaunation, we take centre stage. Because of our movement from supporting actor to the lead role, I think it's important to treat the two events separately. As I said in the introduction, I prefer to think of the megafaunal extinctions as the spark that lit the slow-burning fuse that burnt though the Holocene, only to launch the Anthropocene crisis. In the final chapter, I want to look at what we have learned from the fossil record and see how we can apply that knowledge to the situation we find ourselves in today.

9   Surviving the Anthropocene

> We are not going to be able to operate our Spaceship Earth successfully nor for much longer unless we see it as a whole spaceship and our fate as common. It has to be everybody or nobody.
>
> R. Buckminster Fuller

ARE WE THERE YET?

The increasing level of damage that humans have inflicted on the global environment is reflected in the decision made by geologists to establish a new time interval of geological time: the Anthropocene. Our mega-cities, increasing levels of industrialisation, land clearance, and habitat change will all leave their mark in sediments of Anthropocene age. These sediments will also contain evidence for a massive crash in global diversity. But will any future geologist looking at the sediments deposited during the early Anthropocene regard it as a mass extinction?

The Anthropocene diversity crash is often referred to as the Sixth Extinction, placing it up there with the devastation caused by the Big Five mass extinctions we've been talking about. It's a handy phrase with considerable media impact, but not a term I like using. At a simplistic level, I have argued in this book that there are many more ancient mass extinctions than just five – so calling today's diversity drop the sixth is simply wrong. But there are other reasons for rejecting the term 'the Sixth Extinction'. Firstly, and perhaps most trivially, because the term links today's event with the ancient mass extinctions it makes it sound as if our current crisis is simply part of some natural cycle. This view of history allows some deniers – who may agree that we are living in a mass extinction event – to argue that because there have been at least five big extinctions in the past, today's extinctions are just part of nature. And, since it is obvious that life recovered after every other mass extinction, it will do so again; and therefore we shouldn't worry. This sort of thinking abrogates human culpability in causing today's species loss. I have seen the same argument made against the reality of human-induced climate change. People – unfortunately including some geologists who should know better – argue that the geological record shows that the Earth's climate has been constantly changing in the past, and the current global warming is simply part of a natural cycle. This thinking ignores the fact that the *rate* of climate change is well in excess of anything we've seen in the geological record and leads to the erroneous conclusion that we needn't do anything about it.

However, my strongest reservation about using the Sixth Extinction as a label for what is going on today is that it does not tell the whole story. In Table 1.1, I listed extinction data for some key animal groups. Even given our appalling lack of knowledge of today's biota, the data show that in terms of the number of species actually going extinct, we have not yet reached the extinction level of any of the Big Five: we're not even close. The highest percentage of species that are either extinct or extinct in the wild is recorded by the bivalve molluscs at a low 4%. But, if we expand our view and include species that are under threat of extinction, we get a clearer picture of the biotic crisis we are causing. Indeed, the percentage of extinct and threatened species of bivalves reaches 29%. The only groups that are worse off are the gastropods (33%) and amphibians (34%). If we focus doggedly on the raw extinction levels and ignore the fact that, through our actions, many more species are *under threat* of extinction, then we blind ourselves to the true extent of the biodiversity crisis. We also risk not being able to change anything before it is too late: 'under threat of extinction' species tend to go fully extinct unless things change for the better. For these reasons, when referring to the Anthropocene biodiversity crisis, instead of the Sixth Extinction I prefer to use the term *defaunation*. The term may not be perfect, but it is inclusive – it includes both the number of species going extinct and those under threat of extinction. This gives us a much clearer picture of what we are facing. However, simply because I call it a defaunation rather than the more dramatic Sixth Extinction, you shouldn't think that I am complacent about the situation we are in. In many ways, the Anthropocene defaunation *does* fit some of the criteria we used when defining ancient mass extinctions. It is global in effect, and it is rapid (perhaps more rapid than most of the other extinction events, except the end-Cretaceous event which can be linked to a meteorite impact).

In fact, the changes that we see occurring today – rising temperature due to excess carbon dioxide in the atmosphere coupled with increasingly acidic and oxygen-poor oceans – is a very similar scenario to the one I discussed for the end-Permian mass extinction. In that case, the source of the greenhouse gases was the emplacement of the Siberian Traps. Today, we are the main source of greenhouse gases, largely produced by our burning of fossil fuels. But the results, a warming planet and an increasingly inhospitable ocean, are identical.

In many respects, the situation today is worse than the end-Permian. As a direct result of human activity such as land clearance and habitat change, extinction levels are rising, and the biosphere is becoming less diverse. The damage we are inflicting on the biosphere is critical – we need a healthy, diverse biosphere, because it is a key component of the Earth System. Any damage to the biosphere will result in detrimental changes in the ability of the system to maintain planetary conditions, increasing the likelihood of a mass extinction.

So, perhaps not yet a mass extinction – but there is no doubt that we are on the way. This can be most clearly seen if, instead of looking at absolute extinction values, we compare the rate at which species are going extinct today with the rate they were going extinct before humans had any effect on the environment. Essentially, we want to compare the rate today with that just prior to the earliest Holocene. As you might expect, establishing these rates (particularly the pre-human extinction rate) is difficult – but it has been done.

A pre-human background extinction rate for mammals has been calculated by Anthony Barnosky and his colleagues.[1] They used data from mammals for their study because mammals have an excellent record and they are one of the few animal groups where the level of extinction and the number of species threatened by extinctions has been assessed for most of the group (about 91%, Table 1.1). The result was an estimated pre-human extinction rate of about two extinctions per million species-years (E/MSY).[2] Put simply, this means that, prior to the evolution of humans, if there were a million species of mammals on Earth we could expect, on average, two to go extinct each year owing to natural causes alone. If we had less than a million species alive, we would expect a reduction of the number of species going extinct each year. For example, if there were half a million species alive, then we would expect to see only one species go extinct each year. With a quarter of a million species, we would lose 0.5 species a year (perhaps one every 2 years would be a better way of putting it!). But there are a lot fewer than even a quarter of a million species of mammals alive today. Data from the Red List put the number at about 6,500 species of mammals alive on Earth. This means that on average, without any human interference, the extinction rate would be much less than one species per year: in fact, the number is 0.013 species extinctions per year. To put it another way, this means that if humans hadn't evolved and if the rate of extinction was exactly balanced by the origination of new species – that is the diversity of mammal species stayed constant at about 6,500 – then over the 11,500 years of the Holocene, the planet would have lost and then replaced only about 149 species of mammals.

Next, we need an estimate of the current rate of extinction which, owing to our poor understanding of the state of today's biodiversity, is also very difficult to obtain. However, in 2015, a study

---

[1] See Barnosky et al. (2011) in the Further Reading section.
[2] The actual measurement was 1.8 – but rounding the number up to 2 makes no real difference to the discussion.

led by Gerardo Ceballos[3] produced several estimates of the modern extinction rates using varying criteria. Rather than reviewing them all, I am going to concentrate on the rate they calculated using the number of species of vertebrates (animals with backbones) that the Red List records as having gone extinct or possibly extinct since 1900. Ceballos and his team used vertebrates because, again, most species have been assessed by the Red List. Using their estimate of the modern extinction rate for just the mammals (a subset of vertebrates) and the 2 E/MSY pre-human mammal extinction rate estimated by Barnosky, they concluded that today's extinction rate for mammals is a staggering 55 times higher than the pre-human background rate. And these species are not being replaced at anything like the same rate by newly evolved species.

But because the study looked at all vertebrates, not just the mammals, Ceballos and his colleagues could examine the situation for other subgroups of vertebrates. In order to do this, they had to assume (reasonably, I think) that because mammals are a subset of the vertebrates, the pre-human rate of 2 E/MSY calculated for mammal extinctions could be applied to all of the animals with backbones. If we accept this assumption, their analysis suggests that the current extinction rate for all vertebrates is 53 times the pre-human rate. However, the vertebrate group that fares the worst is the amphibians. The current extinction rate for this group is a heart-breaking 100 times the pre-human background rate. We could quibble about the scale of increase of extinction rates and the assumptions that underlie them, but I don't believe we can argue with the fact that the arrival of humans has caused a massive increase in the rate at which species are going extinct.

Nevertheless, although today's extinction rates are getting comparable to the ancient mass extinctions in terms of species extinctions, I argue that we haven't yet reached a level that I think constitutes a mass extinction. This, however, raises the question:

[3] See Ceballos and others (2015) in the Further Reading section.

how long will it take us to match one of the Big Five extinction events, say a 75% species loss – about the level of the end-Cretaceous event?

We could just extend the current rate of extinction out into the future and see how long it would take. But given the uncertainties around calculating today's rate, that may produce a misleading conclusion. A 2011 study by Barnosky and a team of co-workers approached the problem from a quite different direction.[4] Again, because they offer the best available data, the estimate of the time to a 75% extinction rate is based solely on the vertebrates. As estimates of current extinction rates may be open to criticism, the study calculated two new ones that avoided the problem. The first was the rate that would be needed to cause all the vertebrate species included in the Red List's *threatened* category to go extinct over the next 100 years. The second rate was calculated based on only losing the species that the Red List considers *critically endangered* over 100 years. 'Critically endangered' is a subcategory of 'threatened', so the number of species included will be less, resulting in a lower extinction rate. After calculating these two rates, Barnosky and his colleagues extrapolated them into the future to model what would happen. This provided two estimates of the time it would take for extinctions to reach the 75% species loss, and to say the least, the results are sobering.

The data suggest that if we allow all the currently threatened vertebrate species (the broadest category) to go extinct over the next 100 years and then permit that rate of extinctions to continue with all other vertebrate species, we have between 240 and 540 years left before we reach the point when we have killed off 75% of all vertebrate species. Of course, the biosphere would be in serious trouble well before we reached the 75% point, so we don't want to go there. However, if we choose to make a significant effort to conserve our biota so that only the *critically* endangered subcategory of threatened species go extinct over the next 100 years, and extend that rate

---

[4] See Barnosky and others (2011) in the Further Reading section.

forward, the situation improves. Based on this lower rate of extinctions, the length of time that it will take to reach the 75% extinction level extends out to between 890 and 2,270 years. Yes, the data are for the vertebrates only – but I see no reason why this sort of timing isn't applicable across much of our biota. These estimates of remaining time demonstrate how important conservation efforts are if we are going to avoid sliding blindly into a mass extinction.

To sum up, it is a fact that species are going extinct at a faster rate today than at any other time during the Holocene. The rising extinction rates, the speed of the extinctions, and their global reach all indicate that we are at the beginning of a very significant mass extinction event. If we were to allow this mass extinction to develop fully, it would be unlike any previous extinction events contained in the fossil record. The ancient mass extinction events are the result of natural processes such as meteorite impacts, volcanic eruptions, and sea level changes. Even the megafaunal mass extinction, although human involvement is strongly implied, probably has a significant climatic signal. Today's defaunation, however, is entirely the result of human activity. The planet's biota is in a parlous state, and the estimates given above – while based on limited data and assumptions that could be questioned – clearly indicate where we are headed. It suggests that if we do nothing, we may have as little as 240 years left before we reach a species extinction level that we haven't seen since the end of the Cretaceous. And unlike previous mass extinctions, we have the opportunity *to do something about the situation*. But we have to act rapidly and decisively.

WHAT IF WE DID NOTHING?

What does the fossil record tell us might happen if we didn't act rapidly and decisively, if we just shrugged our shoulders, did nothing, and waited for the mass extinction to run its course? We discussed how life recovers from a mass extinction in Chapter 7, but it's worth summarising that discussion here.

First, there is good news. Life will get through a mass extinction: it always has. The record clearly shows that no matter how devastating the mass extinction, following each event life has recovered. After even the most catastrophic of extinction events, life does rapidly evolve, filling the niches vacated by species that have gone extinct. As a result, the planet's biodiversity will eventually rise and, as the biosphere recovers, the Earth System will reach some sort of equilibrium, though it may not be at the same position as prior to the event. But, as we have seen, that good news comes with a couple of caveats.

The first caveat is that from the perspective of deep time, the fossil record shows that life's recovery from a mass extinction event can be rapid. But that is on the scale of deep time. On a human time scale, recovery will be very slow. We have seen that following the end-Cretaceous event, mammals and plants in Colorado took about a million years to recover. Other estimates suggest that it could take up to 10 million years to regain the diversity lost during a mass extinction. But this is at best an educated guess: it can take much longer. Following the end-Permian mass extinction, there was a period of some 20 million years where no coal deposits were formed. This means that the ecosystems where coal forms (largely swamps and bogs) went missing for over 20 million years. Coral reef ecosystems took even longer. They required an additional 7 million years before they made a reappearance. The results of one study[5] suggest that if we could wind back the current rate of mammal extinctions to pre-human levels, it would take several millions of years for mammal diversity to recover from even today's defaunation.

The second caveat is that we could not be sure what biota will emerge from a mass extinction. The fossil record clearly shows that we cannot predict which animal groups will go extinct or which groups might radiate to fill the empty niches cleared by a mass extinction event. With the data we have available today, we could make

[5] See Davis and others (2018) in the Further Reading section.

some educated guesses. We could use our (admittedly limited) knowledge of the present-day biota to flag which species have a high possibility of going extinct – the Red List threatened category does just that. If we want to avoid the Anthropocene defaunation developing into a full-blown extinction event, we could use these data. Coupling the Red List data with our knowledge of which areas on the planet are being most affected by human activity, we could make an effort to wind back the damage and restore the ecosystem. Recent modelling by a team led by Bernardo Strassburg[6] suggests that if we could conserve the world's remaining natural ecosystems and restore just 30% of the lands we have damaged through clearance, we could reduce the likelihood of species going extinct by about 71%. As a bonus, the regenerating ecosystem would absorb about 465 gigatonnes of carbon dioxide – about 49% of all the gas added to the atmosphere since the industrial revolution. If we make the effort, the rewards will be significant.

If we do nothing, following the mass extinction, we might let evolution do its work without any sort of intervention (which would of course be easy if we ourselves were extinct). In that case, we would have no idea of the biota that would be produced during life's recovery. If we survived, we could, of course, try to manage our way through, replacing dying forests with plants that would be useful to us: fill the African savanna with cattle and sheep to feed us. This approach might lead to lots of individual animals and plants but a very low level of diversity. However, again we should remember that the climatic stability of the planet relies on a *fully operational* Earth System, and that the biosphere is a vitally important component of that system – it's our life support system. The emergence of the modern Earth System some 500 million years or so ago relied on the evolution of complex organisms and the development of a functioning ecosystem. If we could manage the biotic recovery, and it resulted in a lower level of biodiversity, which it almost inevitably would, then the

---

[6] See Strassburg and others (2020) in the Further Reading section.

stability of the Earth System would still be under serious threat. The lack of stability would, in turn, bring with it the possibility of another mass extinction.

I think it's obvious that doing nothing is not an option. And do we really think humanity will be bright enough to manage a successful recovery when we haven't yet been able to manage not to screw things up in the first place? Seems unlikely. Better to sort it now rather than later.

THE RESILIENCE OF THE EARTH SYSTEM

Throughout this book, I have emphasised the importance of the Earth System, the complex network of reservoirs and fluxes that maintains planetary conditions at a climatic equilibrium that is suitable for life. The fossil record shows us that a major shift in the equilibrium point of the Earth System, brought about by environmental forcings such as volcanic eruptions or meteorite impacts, for example, will result in a new climatic setting. This shift in the equilibrium point is often, if not always, accompanied by a mass extinction. Today, human activity is causing major disruptions to many parts of the Earth System. We are applying significant environmental forcings and, as a result, we are risking a shift in the planetary equilibrium point. Such a shift would bring with it a new set of planetary conditions that could persist for millions of years while the system restores some sort of equilibrium. Unless we take urgent measures to reduce the environmental stresses, we are pushing the Earth System towards an equilibrium shift, and a mass extinction would seem inevitable. And as I've said, all the evidence points to it already being under way.

However – and this is very important – it is not too late. As well as containing evidence for ancient equilibrium shifts and mass extinctions, the fossil record shows us just how resilient the Earth System can be. It contains records of times when the Earth System moved very close to an equilibrium shift – but managed to return itself to its original environmental conditions. The best example of this occurred

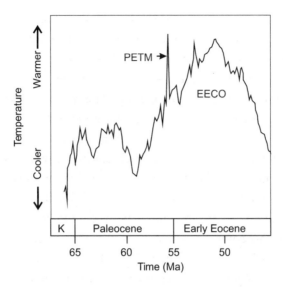

FIGURE 9.1 Global temperature from the late Cretaceous to the early
Eocene, based on deep-sea foraminiferal oxygen isotopes. Time runs along
the horizontal axis, temperature on the vertical. K, Cretaceous; PETM,
Paleocene–Eocene Thermal Maximum; EECO, Early Eocene
Climatic Optimum. Adapted from Wikimedia CC BY-SA 3.0.

some 56 million years ago, at the boundary between the Paleocene
and Eocene epochs.

Figure 9.1 shows the estimated temperature of the planet's
ocean from just before the end-Cretaceous mass extinction through
to the early Eocene. The black line in Figure 9.1 that tracks the
changing temperature is based on the analysis of oxygen isotopes in
the shells of fossil foraminifera. We have met foraminifera several
times earlier in the book, notably when I talked about the causes of
the end-Cretaceous mass extinction. The oxygen isotope data from
foraminiferal shells can be used to give us estimates of sea
water temperatures.

The long-term trend shown in Figure 9.1 indicates that from late
in the Cretaceous onwards the planet's temperature was rising. This
rise in temperatures continued through into the early Paleocene.
Geological analyses of this time interval show that, on average, the

ocean temperature was warmer than today and, although the poles were still colder than the rest of the planet, they were ice-free. Temperature fell back a little in the middle Paleocene, but by its end the trend of rising temperatures had resumed. The oceans reached their warmest level in the early Eocene. This peak in warmth is referred to as the Early Eocene Climatic Optimum or EECO.

Superimposed on this long-term temperature trend are many sharp peaks and troughs recording much shorter-term fluctuations in ocean temperature. Most of these are relatively minor events, but there is one peak in the temperature that towers over all the others. It sits almost exactly at the boundary between the Paleocene and Eocene, and at this point the temperature of the sea increases incredibly quickly, reaching levels that far exceed even the EECO. Some estimates suggest that this peak represents a rise in the temperature of the world's ocean of about 8 degrees Celsius. This highly elevated temperature doesn't last for very long, and equally rapidly it falls back to about the same level as before the rise. This transient climatic event has been named the Paleocene–Eocene Thermal Maximum or PETM.[7]

The rise and fall of the ocean's temperature during the PETM is extremely rapid (in geological terms): the whole event was over in about 200 thousand years. On a human time scale this may seem long, but in terms of deep time it is almost instantaneous. The warming at the PETM has been intensively investigated by scientists who study the Earth's past climate, because it is considered to be a good analogue for today's warming climate. It is being studied to see how the Earth System responds to abrupt warming events.

---

[7] People who study ancient climates *love* acronyms and abbreviations. As well as the EECO and PETM, there are many others, including the LOWE (Late Oligocene Warming Event) and the MMCO (Middle Miocene Climatic Optimum). This latter event is sometimes referred to as the MCO (Miocene Climatic Optimum), not to be confused with the Medieval Climatic Optimum, which is also referred to as the MCO. To stop this confusion, you could refer to the MWP (Medieval Warm Period) or the MCA (Medieval Climatic Anomaly).

We know that today's global warming is a result of the green-house effect: humans are adding ever-increasing amounts of carbon dioxide to the atmosphere, and the result is increasing temperatures. It appears that the PETM is also a result of increased levels of green-house gases in the atmosphere. How do we know that? We cannot directly measure the amount of carbon dioxide in the atmosphere at the time of the PETM, but there is indisputable geochemical evidence that huge amounts of carbon were added to the atmosphere at the same time as the event. Today, it is clear that the carbon dioxide being added to the atmosphere is a result of human activity. When it comes to the PETM, however, there is no consensus as to the source of this greenhouse gas. Volcanic gases and the breakdown of gas hydrates have been blamed, both of which were involved in damaging both the terrestrial climate and oceanic condition at the end of the Permian, triggering a mass extinction (see Figure 6.2). A third possible source of carbon at the PETM, and probably the most likely, is the melting of permafrost in Antarctica. Although it was ice-free at the time, Antarctica would still have been cold enough to have extensive areas of permafrost. The rise in temperature during the late Paleocene leading into the EECO would have caused it to begin melting. Once the organic material contained in the permafrost was exposed to the atmosphere, it would have been oxidised, releasing significant amounts of carbon dioxide and methane. As I said in Chapter 6, methane is a stronger greenhouse gas than carbon dioxide, but fortu-nately it does not stay in the atmosphere for as long. Together these two gases would cause the planet's temperature to rise, which would cause more permafrost to melt, adding still more carbon dioxide and methane to the atmosphere in a positive feedback loop.

The oxidation of the Antarctic permafrost is analogous to our burning coal as a fuel source. When we burn coal, we are taking the ancient carbon that had been safely locked away, converting it to carbon dioxide, and releasing it into the atmosphere. Effectively, we are completing the process of decomposition that started millions of years ago. Frighteningly, today there are areas of significant permafrost melting in many parts of the globe, and this is an additional serious threat to our climate: it could lead to a similar positive feedback loop to the one mentioned above, potentially leading to runaway climate change.

At the PETM, the greenhouse-gas-induced temperature rise resulted in significant changes to the Earth's biota. As the planet warmed, the tropics expanded, and warmth-loving plants and animals migrated towards the poles, causing a radical rearrangement of the planet's biogeography. In the ocean, there were some areas of lowered oxygen and others of raised acidity. At the surface of the ocean, numbers of dinoflagellates, a form of algae, exploded. Today, blooms of dinoflagellates cause red tides (and many cases of shellfish poisoning). However, despite all the climatic changes, there was no mass extinction. There was only one fossil group that showed any significant level of extinction: the foraminifera. Even then, it wasn't the whole foraminifera group that suffered, just those that live on the ocean floor. Planktic foraminifera – those that live their lives floating in the water column – were largely unaffected.[8]

It seems that, at the PETM a sudden influx of greenhouse gases triggered an episode of global warming. This started to push the Earth System out of its equilibrium state towards a tipping point. If it had pushed through that tipping point, then the system would have undergone an equilibrium shift, possibly accompanied by a descent into a full-scale mass extinction. But it didn't get there. Once the source of the carbon dioxide and methane was depleted, the various feedback cycles that link all the components of the Earth System were able to strip the excess greenhouse gases out of the atmosphere quickly enough to avoid the shift in the position of the equilibrium and avoid a mass extinction.

---

[8] I could speculate here that perhaps the reason there was no mass extinction is that there was no long-term environmental stress predisposing the biota to a mass extinction such as we saw in the end-Permian and end-Cretaceous cases.

The biosphere would have been heavily involved. On land, plants, encouraged by the high levels of carbon dioxide, would increase in numbers and grow faster, taking carbon out of the atmosphere in the process. In the ocean, the single-celled coccolithospheres that use carbon when building their shells would have responded to the increased levels of carbon dioxide by blooming: millions upon millions of minute creatures absorbing carbon dioxide from the atmosphere and using the carbon to produce their tiny calcium carbonate shells. On their death, the shells would fall to the ocean floor, locking away ever-increasing amounts of carbon as deposits of chalk. The ocean itself would dissolve carbon dioxide in its surface waters. In the geosphere, the level of chemical weathering of rocks, which occurs in warmer tropical areas, would increase as the tropics expanded on a warming planet. We have already seen that chemical weathering is a very effective carbon sink, capable of removing vast amounts of carbon dioxide from the atmosphere. These natural processes operating across the entire Earth System managed to return the system to its previous state, avoiding a mass extinction.

There is much that we can learn from the parallels between the PETM and the situation we find ourselves in today. But the hopeful message I take from the event is that the Earth System is inherently resilient. The processes that saved the planet from a mass extinction at the PETM still operate today – but we have to give them a chance. We must allow them to function properly in order to pull us back from the brink. Firstly, we must ensure that the environmental forcings that are perturbing the Earth System are significantly reduced. The key here is to bring climate change under control by greatly reducing our emissions. At the same time, for the system's feedback loops to operate efficiently, we need a fully functioning, diverse biosphere. We must allow our biosphere to heal and at least maintain our current level of biodiversity by stepping up our conservation efforts. One clear and achievable goal would be simply to fully document our biota to help us make environmentally sensible decisions.

LIVING AND DYING IN THE MARGIN OF ERROR

The estimated 8.7 million species alive on Earth today are just the slightest of twigs on the furthest reaches of the one true tree of life. These twigs can be traced back, through ever-thickening branches, right down to the roots of the tree some 3.7 billion years ago. The sheer number of species that have lived and died out over that immense period dwarfs the current biodiversity. Today, all living things exist within the margin of error of any estimate of the total number of species that have ever existed on Earth. The scrap of biodiversity that remains after 3.7 billion years of evolution now stands on a knife's edge. In this book, I have argued that to fully understand the current defaunation and to look for ways we can mitigate or reverse the situation, it is important to set the event into its historical context.

Because of the constraints imposed by the fossil record, we started our detailed look at the history of life a long way up from its roots, around 700 million years ago. By this stage, about 80% of life's history is already over. The fossil record improves from that point onwards, and rocks start to record enough physical and geochemical evidence to broadly set out what was happening. Of particular importance to the story of life on Earth is the history of oxygen. Absent on the early Earth, oxygen became a permanent component of the atmosphere about 2.45 billion years ago. From that point onwards, hesitantly, oxygen levels in the atmosphere started to rise. Eventually, even the deepest parts of the ocean became oxygenated. The increasing levels of oxygen allowed life to make the transition from single cells to complex multicellular animals, bringing with them the first modern style of ecosystems. Together, the appearance of large multicellular animals and complex ecosystems also heralded the emergence of the modern Earth System.

Evolution links all species on Earth. Every species alive today could, if we had a perfect fossil record and continuous sampling, have its ancestry traced back to our last common ancestor some 3.7 billion years ago, at the root of the tree of life. In turn, the Earth System links all living things with the Earth's physical processes. Life in all its diversity, the air we breathe, the rocks beneath our feet, and the water that fills our oceans are all intimately connected. A change or disruption in any of these components of the Earth System will be felt and responded to by all the others.

Following the evolution of multicellular life, about 500 million years ago the diversity of life exploded. New species evolved at a rate that has not been matched since. The improvement in the fossil record that accompanied the Cambrian Explosion allowed palaeontologists to examine, in detail, the rollercoaster ride that diversity has taken up to the present day. And what a ride it was, full of rapid increases in biodiversity and precipitous falls due to mass extinctions. It's clear that there is continuing discussion about the exact path that biodiversity followed, but additional research is very likely to sort these differences out. Despite the disagreements, there is something I think everybody agrees on, and that is the importance of mass extinctions in the history of the biosphere. Mass extinctions are recognised as massive drops in biodiversity. The Big Five mass extinctions, for example, may have resulted in extinctions of marine species of between 75% and 96%. But mass extinctions do more than simply reduce the planet's diversity. The clearing of environmental niches and their subsequent refilling by a different set of species gives mass extinctions the ability to reconstruct entire biotas and makes them a powerful instrument of biological change. They both shape the overall diversity of the planet and drive major restructurings of its biota.

Mass extinctions are linked with major shifts in the climatic equilibrium that is maintained by the Earth System. Unfortunately, through our disturbance of the planet's ecosystem, the Earth is losing species at an ever-increasing rate, and an equilibrium shift is threatened, putting a significant proportion of the human population at risk (potentially of total extinction) and at the very least lowering the quality of life for the remainder. This is happening because we are damaging the biosphere, one of the key components of the Earth

System, both directly – as we clear the land to feed and house our growing population – and indirectly through climate change, a result of our increasing levels of greenhouse gas emissions through the burning of fossil fuels. Modern data imply that we have not yet reached the level of species loss we would expect to record in a mass extinction. But we are causing a significant defaunation. Analysis of the fossil record shows that we still have some time to change our behaviour if we want to avoid a descent into another full-blown mass extinction. But it's not a lot of time. The better the decisions we make now in order to safeguard the planet's biodiversity and reduce greenhouse emissions, the more time we gain and the more chance we have to push back calamity.

Equally, the fossil record shows us what will happen if we do nothing. We will slide into a new set of climatic conditions and enter a mass extinction where we could lose most of the species alive today. If humans were to survive this mass extinction, we would end up living on a climatically unstable planet with only a small percentage of the biodiversity we see around us today. I cannot see how we could manage our way through this sort of ecological upheaval – so we must act to avoid it.

Appendix 1: The Geological Time Scale

I have included here the latest version of the complete international Geological Time Scale, reproduced with the permission of the International Commission on Stratigraphy. Here it's reproduced in greyscale, but to see it in all its colourful glory, go to: www.stratigraphy.org/ICSchart/ChronostratChart2020-03.pdf

It really is a thing of beauty.

For the sake of clarity, throughout this book I have referred to this chart as the Geological Time Scale. More properly, it should be called a chronostratigraphic chart, not a time scale. The reason for this is the way geologists separate time and rock units, as I discussed in Chapter 2.

You should also be aware that in order to fit the entire diagram on one page, the millions of years are not presented in a linear fashion. For example, the Precambrian column on the far right represents something like 4,000 million years, but the entire remainder of the diagram represents only (!) 540 million years. The Quaternary at the top of the left-hand column is about 2.6 million years long. The underlying unit, the Neogene, which on paper looks about the same thickness, in fact is about 20 million years long. As reproduced here, the Quaternary is about 2 centimetres in length. To reproduce the entire chart in a single column at the same scale would require a piece of paper about 35 metres long.

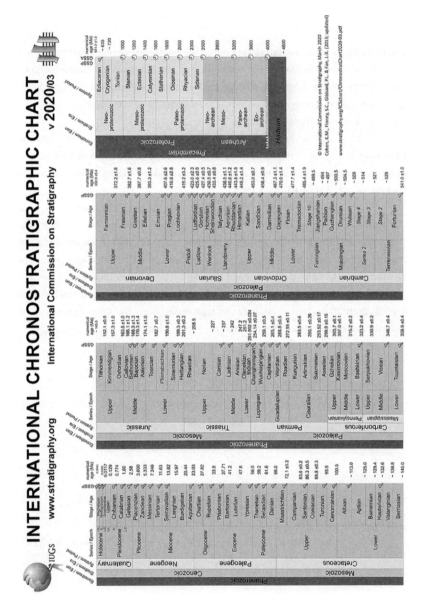

FIGURE A.I

Appendix 2: How Long Will It Take to Count to 1 Million? A Rough Guide

One of my first teaching tasks of the year is to introduce the important concept of geological time to a first-year class. This is quite a challenge; I'm not sure that I can entirely get my head around all those millions of years myself, so how can I convince the first years to think beyond the day-by-day human time scale and embrace deep time? But geology just does not make sense without some under-standing of deep time, so I have to try.

My initial approach was to compare the human and geological time scales and impress on the students the vastness of deep time. To illustrate the human time frame, I used to read poetry to them – to be specific, Andrew Marvell's *To His Coy Mistress*. The poem is firmly fixed in a very human time scale (and involves sex, which I thought they'd like) but I soon discovered that first years weren't keen on the metaphysical poets – the look of boredom on 350 faces was a real low point in my lecturing career. So instead I now use a slide with lots of images of various events on it shown in order of increasing age (a recent earthquake, political events, my favourite 1960 prog rock group,[1] publication of the *Origin of Species*, all the way back to ancient Egypt and cave paintings), and I ask them to let me know when we reach an image that they consider 'old'. We don't usually get too far past the First World War.

Deep time is difficult – we can't directly experience it, so I decided that a more direct, intellectual approach was needed. I start by telling them the age of the Universe (about 14 billion years) and the age of the Earth (about 4.7 billion years), and point out that the country they live in, Aotearoa/New Zealand, being only about 500 million years old, is a relative youngster. By the time that Zealandia started to form, almost 90% of the Earth's history was over. I let the figures sink in. Then I ask, 'How long will it take me to count to 1 million?' I get a few estimates (a million seconds is a common one), then I give them my estimate.

These are the mechanics I followed to make my estimate:

- I timed myself reading out three blocks of twenty numbers from between 100 and 1,000,000. I avoided the first 100 numbers because reading out 1, 2, 3, 4, etc. is

[1] Jethro Tull.

quicker than all those long numbers, and including the lower numbers could have biased the result. I endeavoured to maintain the same reading pace at all times.

- For each block of numbers, I repeated the count three times, then averaged the results. I then averaged the averages for each block to come up with a final reading time for a block of twenty numbers.
- I then multiplied that time by the number of blocks of twenty numbers in 1 million, and hey presto, it's done (of course I divided by 60 to get minutes, then by 60 again to get hours, then 24 to get days).

Sitting in a quiet room (except for the muffled laughter from my wife), I started chanting the numbers, feeling a bit like some old Benedictine monk reciting the daily lesson. I quickly discovered how bloody difficult it is to read out blocks of long numbers without losing your place. I had so many false starts that it took far longer than I expected. My recommendation is not to try it – I have suffered for the sake of science so you don't have to.

So, what is the result? How long will it take to count to 1 million?

On average, each block of numbers took about 57 seconds to count (the individual times were very consistent at around 57 seconds with one outlying set that took 60). Simple arithmetic suggests that with 50,000 blocks of twenty numbers in 1 million it would take 2,850,000 seconds to do the count, about 47,500 minutes. That's around 792 hours or 33 days – more than a month.

So, it would take over a month to count to 1 million (no sleeping and no meals or toilet breaks). It takes a mere 3,500 months (about 290 years) of counting to reach the time that life first appeared on the planet. About 4,500 months (about 375 years) of counting gets you to the age of the Earth. You would have to chant away for about 14,000 months (nearly 1,200 years) to reach the age of the Universe.

Does that give you at least some sense of the scale of deep time? Hard, isn't it?

Further Reading

Included here are the main readings I found useful in shaping my thinking on the various topics covered in this book. For easy access, they are arranged chapter by chapter.

INTRODUCTION

Mora, C., D. P. Tittensor, S. Adl, A. G. B. Simpson & B. Worm (2011). How many species are there on Earth and in the ocean? *PLoS Biology* **9**(8) e1001127.

THE ANTHROPOCENE AND THE EARTH SYSTEM

Berentson, Q. (2012). *Moa*. Nelson, New Zealand: Craig Potten Publishing. (Book)

Braje, T. J. (2015). Earth systems, human agency, and the Anthropocene: Planet Earth in the Human Age. *Journal of Archaeological Research*, **23**(4) 369–396.

Brook, B. W., N. S. Sodhi & C. J. A. Bradshaw (2008). Synergies among extinction drivers under global change. *Trends in Ecology and Evolution*, **23**(8) 453–460.

Budiansky, S. (1994). Extinction or miscalculation? *Nature*, **370**(6485) 104–105.

Ceballos, G. & P. R. Ehrlich (2018). The misunderstood sixth mass extinction. *Science*, **360**(6393) 1080–1081.

Ceballos, G., P. R. Ehrlich & R. Dirzo (2017). Biological annihilation via the ongoing sixth mass extinction signalled by vertebrate population losses and declines. *Proceedings of the National Academy of Sciences USA*, **114**(30) E6089–E6096.

Dirzo, R. & P. H. Raven (2003). Global state of biodiversity and loss. *Annual Review of Environment and Resources*, **28**(1) 137–167.

Dirzo, R., H. S. Young, M. Galetti, et al. (2014). Defaunation in the Anthropocene. *Science*, **345**(6195) 401–406.

Duffy, J. E. (2009). Why biodiversity is important to the functioning of real-world ecosystems. *Frontiers in Ecology and the Environment*, **7**(8) 437–444.

IUCN 2020. The IUCN Red List of Threatened Species. Version 2020-2. https://www.iucnredlist.org (Accessed 18 September 2020.)

Kolbert, E. (2014). *The Sixth Extinction: An Unnatural History*. New York: Henry Holt and Company LLC. (Book)

Lenton, T. M., S. J. Daines, J. G. Dyke, et al. (2018). Selection for Gaia across multiple scales. *Trends in Ecology and Evolution*, 33(8) 336–345.

Pievani, T. (2014). The sixth mass extinction: Anthropocene and the human impact on biodiversity. *Rendiconti Lincei*, 25(1) 85–93.

Pimm, S. L., C. N. Jenkins, R. Abell, et al. (2014). The biodiversity of species and their rates of extinction, distribution, and protection. *Science*, 344(6187) 246752.

Pimm, S. L. & P. Raven (2000). Extinction by numbers. *Nature*, 403(6772) 843–845.

Pimm, S. L., G. J. Russell, J. L. Gittleman & T. M. Brooks (1995). The future of biodiversity. *Science*, 269(5222) 347–350.

Plotnick, R. E. & K. A. Koy (2020). The Anthropocene fossil record of terrestrial mammals. *Anthropocene*, 29 1–15.

Pryon, R. A. (2017). We don't need to save endangered species. Extinction is part of evolution. https://www.washingtonpost.com/outlook/we-dont-need-to-save-endangered-species-extinction-is-part-of-evolution/2017/11/21/57fc5658-cdb4-11e7-a1a3-0d1e45a6de3d_story.html? utm_term=.38245303c44c

Rollinson, H. (2007). *Early Earth Systems*. Malden, MA: Blackwell Publishing. (Book)

Steffen, W., J. Rockström, K. Richardson, et al. (2018). Trajectories of the Earth System in the Anthropocene. *Proceedings of the National Academy of Sciences USA*, 115(33) 8252–8259.

Tyrrell, T. (2013). *On Gaia*. New Jersey: Princeton University Press. (Book)

Waters, C. N., J. Zalasiewicza, C. Summerhayes, et al. (2018). Global Boundary Stratotype Section and Point (GSSP) for the Anthropocene series: Where and how to look for potential candidates. *Earth-Science Reviews*, 178 379–429.

Waters, C. N., J. A. Zalasiewicz, M. Williams, M. A. Ellis & A. M. Snelling (2014). A stratigraphical basis for the Anthropocene? In C. N. Waters, et al. (Eds.), *A Stratigraphical Basis for the Anthropocene. Geological Society Special Publication* 395, 1–21. London: Geological Society.

Winfree, R., J. R. Reilly, I. Bartomeus, et al. (2018). Species turnover promotes the importance of bee diversity for crop pollination at regional scales. *Science*, 359 (6377) 791–793.

Young, H. S., D. J. McCauley, M. Galetti, et al. (2016). Patterns, causes, and consequences of Anthropocene defaunation. *Annual Review of Ecology, Evolution and Systematics*, 47(1) 333–358.

Zalasiewicz, J., C. N. Waters, M. Williams, et al. (2015). When did the Anthropocene begin? A mid-twentieth century boundary level is stratigraphically optimal. *Quaternary International*, 383 196–203.

A SHORT DETOUR: THE FOSSIL RECORD AND THE GEOLOGICAL TIME SCALE

Cohen, K. M., S. C. Finney, P. L. Gibbard & J.-X. Fan (2013; updated). The ICS International Chronostratigraphic Chart. *Episodes*, **36** 199–204.

Coyne, J. A. (2009). *Why Evolution is True*. Oxford: Oxford University Press. (Book)

Foote, M. & A. L. Miller (2007). *Principles of Paleontology* 3rd Edition. New York: W. H. Freeman and Company. (Book)

Mora, C., D. P. Tittensor, S. M. Adl & A. Simpson (2011). How many species are there on Earth and in the ocean? *PLoS Biology*, **9**(8) e1001127.

Raup, D. M. & J. J. Sepkoski (1982). Mass extinctions in the marine fossil record. *Science*, **215**(4539) 1501–1503.

Rudwick, M. J. S. (2005). *Bursting the Limits of Time: The Reconstruction of Geohistory in the Age of Revolution*. Chicago: Chicago University Press. (Book)

Rudwick, M. J. S. (2008). *Worlds Before Adam: The Reconstruction of Geohistory in the Age of Reform*. Chicago: Chicago University Press. (Book)

THE ORIGIN OF ANIMALS AND THE EMERGENCE OF THE EARTH SYSTEM

Anon. (2011). *Geology of the Flinders Ranges National Park*. Adelaide: Government of South Australia.

Bellefroid, E. J., A. v. S. Hood, P. F. Hoffman, et al. (2018). Constraints on Paleoproterozoic atmospheric oxygen levels. *Proceedings of the National Academy of Sciences*, https://doi.org/10.1073/pnas.1806216115

Blamey, N. J. F., U. Brand, J. Parnell, et al. (2016). Paradigm shift in determining Neoproterozoic atmospheric oxygen. *Geology*, **44**(8) 651–654.

Bobrovskiy, I., J. M. Hope, A. Ivantsov, et al. (2018). Ancient steroids establish the Ediacaran fossil *Dickinsonia* as one of the earliest animals. *Science*, **361**(6408) 1246–1249.

Brasier, M. (2009). *Darwin's Lost World*. Oxford/New York: Oxford University Press. (Book)

Briggs, D. E. G. (2015). The Cambrian explosion. *Current Biology*, **25**(19) R864–R868.

Brocks, J. J., A. J. M. Jarrett, E. Sirantoine, et al. (2017). The rise of algae in Cryogenian oceans and the emergence of animals. *Nature*, **548**(7669) 578–581.

Butterfield, N. J. (2011). Animals and the invention of the Phanerozoic Earth system. *Trends in Ecology and Evolution*, **26**(2) 81–87.

Canfield, D. E. (2014). *Oxygen: A Four Billion Year History*. New Jersey: Princeton University Press. (Book)

Canfield, D. E., S. W. Poulton & G. M. Narbonne (2007). Late-Neoproterozoic deep-ocean oxygenation and the rise of animal life. *Science*, **315**(5808) 92–95.

Canfield, D. E., L. Ngombi-Pemba, E. U. Hammarlund, et al. (2013). Oxygen dynamics in the aftermath of the Great Oxidation of Earth's atmosphere. *Proceedings of the National Academy of Sciences USA*, **110**(42) 16736–16741.

Cao, X. & H. Bao (2013). Dynamic model constraints on oxygen-17 depletion in atmospheric O_2 after a snowball Earth. *Proceedings of the National Academy of Sciences USA*, **110**(36) 14546–14550.

Carbone, C. & G. M. Narbonne (2014). When life got smart: The evolution of behavioral complexity through the Ediacaran and Early Cambrian of NW Canada. *Journal of Paleontology*, **88**(2), 309–330.

Chen, J.-Y., D. J. Bottjer, E. H. Davidson, et al. (2009). Phase contrast synchrotron X-ray microtomography of Ediacaran (Doushantuo) metazoan microfossils: Phylogenetic diversity and evolutionary implications. *Precambrian Research*, **173**(1) 191–200.

Crockford, P. W., J. A. Hayles, H. Bao, et al. (2018). Triple oxygen isotope evidence for limited mid-Proterozoic primary productivity. *Nature*, **559**(7715) 613–616.

Daley, A. C., J. B. Antcliffe, H. B. Drage & S. Pates (2018). Early fossil record of Euarthropoda and the Cambrian Explosion. *Proceedings of the National Academy of Sciences USA*, **115**(21) 5323–5331.

Darroch, S. A. F., M. Laflamme & P. J. Wagner (2018). High ecological complexity in benthic Ediacaran communities. *Nature Ecology and Evolution*, **2**(10) 1541–1547.

Darroch, S. A. F., E. F. Smith, M. Laflamme & D. H. Erwin (2018). Ediacaran extinction and Cambrian Explosion. *Trends in Ecology and Evolution*, **33**(9) 653–663.

Darroch, S. A. F., E. A. Sperling, T. H. Boag, et al. (2015). Biotic replacement and mass extinction of the Ediacara biota. *Proceedings of the Royal Society B: Biological Sciences*, **282**(1814) 20151003.

Droser, M. L. & J. G. Gehling (2015). The advent of animals: The view from the Ediacaran. *Proceedings of the National Academy of Sciences USA*, **112**(16) 4865–4870.

Droser, M. L., L. G. Tarhan & J. G. Gehling (2017). The rise of animals in a changing environment: Global ecological innovation in the Late Ediacaran. *Annual Review of Earth and Planetary Sciences*, **45**(1) 593–617.

Eickmann, B., A. Hofmann, M. Wille, et al. (2018). Isotopic evidence for oxygenated Mesoarchaean shallow oceans. *Nature Geoscience*, **11**(2) 133–138.

Erwin, D. H. & J. W. Valentine (2013). *The Cambrian Explosion: The Construction of Animal Biodiversity*. Greenwood Village, CO: Roberts and Company. (Book)

Evans, S. D., I. V. Hughes, J. G. Gehling & M. L. Droser, (2020). Discovery of the oldest bilaterian from the Ediacaran of South Australia. *Proceedings of the National Academy of Sciences USA*, **117**(14) 7845–7850.

Fakhraee, M., Crowe, S. A., & Katsev, S. (2018). Sedimentary sulfur isotopes and Neoarchean ocean oxygenation. *Science Advances*, *4*(1), e1701835. doi:10.1126/sciadv.1701835.

Gehling, J. G., J. B. Jago, J. R. Paterson, G. A. Brock, M. L. Droser, (2012). *Field Trip S-4 Ediacaran–Cambrian of South Australia*. 34th International Geological Congress, Brisbane, Australia.

Gehling, J. G. & M. Droser (2012). Ediacarian stratigraphy and the biota of the Adelaide Geosyncline, South Australia. *Episodes*, **35**(1) 236–246.

Guilbaud, R., B. J. Slater, S. W. Poulton, et al. (2018). Oxygen minimum zones in the early Cambrian ocean. *Geochemical Perspectives Letters*, **6** 33–38.

Holland, H. D. (2009). Why the atmosphere became oxygenated: A proposal. *Geochimica et Cosmochimica Acta*, **73**(18) 5241–5255.

Jago, J. B., J. G. Gehling, J. R. Paterson, et al. (2012). Cambrian stratigraphy and biostratigraphy of the Flinders Ranges and the north coast of Kangaroo Island, South Australia. *Episodes*, **35**(1) 247–255.

Kaufman, A. J. (2014). Early Earth: Cyanobacteria at work. *Nature Geoscience*, **7**(4) 253–254.

Knoll, A. H. (2003). *Life on a Young Planet. The First Three Billion Years of Evolution on Earth*. Princeton New Jersey: Princeton University Press. (Book)

Koehler, M. C., R. Buick, M. A. Kipp, E. E. Stüeken & J. Zaloumis (2018). Transient surface ocean oxygenation recorded in the ~2.66-Ga Jeerinah Formation, Australia. *Proceedings of the National Academy of Sciences USA*, **115**(30) 7711–7716.

Laflamme, M. (2010). Wringing out the oldest sponges. *Nature Geoscience*, **3**(9) 597–598.

Lenton, T. M., R. A. Boyle, S. W. Poulton, et al. (2014). Co-evolution of eukaryotes and ocean oxygenation in the Neoproterozoic era. *Nature Geoscience*, **7**(4) 257–265.

Li, Z.-Q., L.-C. Zhang, C.-J. Xue, et al. (2018). Earth's youngest banded iron formation implies ferruginous conditions in the Early Cambrian ocean. *Scientific Reports*, **8**(1) 9970.

Maloof, A. C., C. V. Rose, R. Beach, et al. (2010). Possible animal-body fossils in pre-Marinoan limestones from South Australia. *Nature Geoscience*, **3**(9) 653–659.

Marshall, C. R. (2006). Explaining the Cambrian 'explosion' of animals. *Annual Review of Earth and Planetary Sciences*, **34**(1) 355–384.

McMenamin, M. A. (1998). *The Garden of Ediacara: Discovering the First Complex Life*. New York: Colombia University Press. (Book)

Meysman, F. J. R. (2014). Biogeochemistry: Oxygen burrowed away. *Nature Geoscience*, 7(9) 620–621.

Mills, B., T. M. Lenton & A. J. Watson. (2014). Proterozoic oxygen rise linked to shifting balance between seafloor and terrestrial weathering. *Proceedings of the National Academy of Sciences USA*, 111(25) 9073–9078.

Mills, D. B., L. M. Ward, C. Jones, et al. (2014). Oxygen requirements of the earliest animals. *Proceedings of the National Academy of Sciences USA*, 111(11) 4168–4172.

Muscente, A. D., T. H. Boag, N. Bykova & J. D. Schiffbauer (2018). Environmental disturbance, resource availability, and biologic turnover at the dawn of animal life. *Earth-Science Reviews*, 177 248–264.

Planavsky, N. J., C. T. Reinhard, X. Wang, et al. (2014). Low Mid-Proterozoic atmospheric oxygen levels and the delayed rise of animals. *Science*, 346(6209) 635–638.

Sahoo, S. K., N. J. Planavsky, G. Jiang, et al. (2016). Oceanic oxygenation events in the anoxic Ediacaran ocean. *Geobiology*, 14(5) 457–468.

Schiffbauer, J. D. & S. Xiao (2014). An examination of life history and behavioral evolution across the Ediacaran–Cambrian transition. *Journal of Paleontology*, 88(2) 205–206.

Schopf, J. W., K. Kitajima, M. J. Spicuzza, A. B. Kudryavtsev & J. W. Valley (2018). SIMS analyses of the oldest known assemblage of microfossils document their taxon-correlated carbon isotope compositions. *Proceedings of the National Academy of Sciences USA*, 115(1) 53–58.

Seilacher, A. (1989). Vendozoa: Organismic construction in the Proterozoic biosphere. *Lethaia*, 22(3) 229–239.

Sperling, E. A., C. A. Frieder, A. V. Raman, et al. (2013). Oxygen, ecology, and the Cambrian radiation of animals. *Proceedings of the National Academy of Sciences USA*, 110(33) 13446.

van de Velde, S., B. Mills, F. J. R. Meysman & T. M. Lenton (2018). Early Palaeozoic ocean anoxia and global warming driven by the evolution of shallow burrowing. *Nature Communications*, 9(1) 2554.

Wei, G.-Y., N. J. Planavsky, L. G. Tarhan & X. Chen (2018). Marine redox fluctuation as a potential trigger for the Cambrian Explosion. *Geology*, 46(7) 587–590.

Wen, H., J. Carignan, Y. Zhang, et al. (2011). Molybdenum isotopic records across the Precambrian–Cambrian boundary. *Geology*, 39(8) 775–778.

Wood, R., A. G. Liu, F. Bowyer, et al. (2019). Integrated records of environmental change and evolution challenge the Cambrian explosion. *Nature Ecology and Evolution*, 3(4) 528–538.

Zhang, F., S. Xiao, B. Kendall, et al. (2018). Extensive marine anoxia during the terminal Ediacaran Period. *Science Advances*, 4(6) eaan8983.

Zhu, M., A. Yu. Zhuravlev, R. A. Wood, F. Zhao & S. S. Sukhov (2017). A deep root for the Cambrian explosion: Implications of new bio- and chemostratigraphy from the Siberian Platform. *Geology*, **45**(5) 459–462.

DOCUMENTING ANCIENT BIODIVERSITY

Alroy, J. (2010a). Geographical, environmental and intrinsic biotic controls on Phanerozoic marine diversification. *Palaeontology*, **53**(6) 1211–1235.

Alroy, J. (2010b). The shifting balance of diversity among major marine animal groups. *Science*, **329**(5996) 1191–1194.

Alroy, J., M. Aberhan, D. J. Bottjer, et al. (2008). Phanerozoic trends in the global diversity of marine invertebrates. *Science*, **321**(5885) 97–100.

Benton, M. (1995). Diversification and extinction in the history of life. *Science*, **268** (5207) 52–58.

Benton, M. J. (2001). Biodiversity on land and in the sea. *Geological Journal*, **36**(3–4) 211–230.

Benton, M. J. (2013). Origins of biodiversity. *Palaeontology*, **56**(1) 1–7.

Benton, M. J. & B. C. Emerson (2007). How did life become so diverse? The dynamics of diversification according to the fossil record and molecular phylogenetics. *Palaeontology*, **50**(1) 23–40.

Fan, J.-X., S.-Z. Shen, D. H. Erwin, et al. (2020). A high-resolution summary of Cambrian to Early Triassic marine invertebrate biodiversity. *Science*, **367**(6475) 272–277.

Gould, S. J. (1989). *Wonderful Life: The Burgess Shale and the Nature of History*. New York: W. W. Norton. (book)

Harper, D. A. T. & M. J. Benton (2001). Preface: History of biodiversity. *Geological Journal*, **36**(3–4) 185–186.

Jablonski, D., K. Roy, J. W. Valentine, R. M. Price & P. S. Anderson (2003). The impact of the Pull of the Recent on the history of marine diversity. *Science*, **300** (5622) 1133–1135.

Phillips, J. (1860). *Life on the Earth: Its Origin and Succession*. Cambridge: Macmillan and Co. (Book)

Sepkoski, J. J. (1981). A factor analytic description of the Phanerozoic marine fossil record. *Paleobiology*, **7**(1) 36–53.

Sepkoski, J. J. (1984). A kinetic model of Phanerozoic taxonomic diversity. III. Post-Paleozoic families and mass extinctions. *Paleobiology*, **10**(2) 246–267.

Sepkoski, J. J. (1993). Ten years in the library: New data confirm paleontological patterns. *Paleobiology*, **19**(1) 43–51.

Winchester, S. (2001). *The Map That Changed the World: William Smith and the Birth of Modern Geology*. New York: Harper Collins. (Book)

Womack, T. M., J. S. Crampton & M. J. Hannah (2020). The Pull of the Recent revisited: Negligible species-level effect in a regional marine fossil record. *Paleobiology*, **46**(4), 470–477.

MASS EXTINCTIONS: THE BASICS

Alroy, J. (2008). Dynamics of origination and extinction in the marine fossil record. *Proceedings of the National Academy of Sciences USA*, **105** (Supplement 1) 11536–11542.

Alroy, J., M. Aberhan, D. J. Bottjer, et al. (2008). Phanerozoic trends in the global diversity of marine invertebrates. *Science*, **321**(5885) 97–100.

Alroy, J., C. R. Marshall, R. K. Bambach, et al. (2001). Effects of sampling standardization on estimates of Phanerozoic marine diversification. *Proceedings of the National Academy of Sciences USA*, **98**(11) 6261–6266.

Alvarez, L. W., W. Alvarez, F. Asaro & H. V. Michel (1980). Extraterrestrial cause for the Cretaceous–Tertiary extinction. *Science*, **208**(4448) 1095–1108.

Alvarez, W. (2003). Comparing the evidence relevant to impact and flood basalt at times of major mass extinctions. *Astrobiology*, 3(1) 153–161.

Alvarez, W. & R. A. Muller (1984). Evidence from crater ages for periodic impacts on the Earth. *Nature*, **308**(5961) 718–720.

Arens, N. C. & I. D. West (2008). Press-pulse: A general theory of mass extinction? *Paleobiology*, **34**(4) 456–471.

Artemieva, N. & J. Morgan (2020). Global K–Pg layer deposited from a dust cloud. *Geophysical Research Letters*, **47**(6) https://doi.org/10.1029/2019gl086562

Bailer-Jones, C. A. L. (2009). The evidence for and against astronomical impacts on climate change and mass extinctions: A review. *International Journal of Astrobiology*, **8**(3) 213–219.

Bailer-Jones, C. A. L. (2011). Bayesian time series analysis of terrestrial impact cratering. *Monthly Notices of the Royal Astronomical Society*, **416**(2) 1163–1180.

Bambach, R. K. (2006). Phanerozoic biodiversity mass extinctions. *Annual Review of Earth and Planetary Sciences*, **34**(1) 127–155.

Bambach, R. K., A. H. Knoll & S. C. Wang (2004). Origination, extinction, and mass depletions of marine diversity. *Paleobiology*, **30**(4) 522–542.

Benton, M. J. (2003). *When Life Nearly Died*. London: Thames and Hudson. (Book)

Berner, R. A. & D. J. Beerling (2007). Volcanic degassing necessary to produce a $CaCO_3$ undersaturated ocean at the Triassic–Jurassic boundary. *Palaeogeography, Palaeoclimatology, Palaeoecology*, **244**(1–4) 368–373.

Brenchley, P. J., J. D. Marshall & C. J. Underwood. (2001). Do all mass extinctions represent an ecological crisis? Evidence from the Late Ordovician. *Geological Journal*, **36**(3–4) 329–340.

Cloud, P. E. (1948). Some problems and patterns of evolution exemplified by fossil invertebrates. *Evolution*, **2**(4) 322–350.

Davis, M., P. Hut & R. A. Muller (1984). Extinction of species by periodic comet showers. *Nature*, **308**(5961) 715–717.

Editorial (2013). The upside of impacts. *Nature Geoscience*, **6**(12) 987.

Fischer, A. G. & M. A. Arthur (1977). Secular variations in the pelagic realm. *Society of Economic Paleontologists and Mineralogists Special Publication*, **25** 19–50

Fisher, A. (1985). Death star. *Popular Science*, **226** 72–77.

Hatfield, C. B. & M. J. Camp (1970). Mass extinctions correlated with periodic galactic events. *GSA Bulletin*, **81**(3) 911–914.

Kornei, K. (2018). Huge global tsunami followed dinosaur-killing asteroid impact. *EOS*, **99** https://doi.org/10.1029/2018EO112419

Jablonski, D., (1994). Extinctions in the fossil record. *Philosophical Transactions: Biological Sciences*, **344**(1307) 11–17.

Lipowski, A. (2005). Periodicity of mass extinctions without an extraterrestrial cause. *Physical Review E*, **71**(5) 052902.

MacLeod, N. (2003). Causes of Phanerozoic extinctions. In L. Rothschild & A. Lister (Eds.), *Evolution on Planet Earth* (pp. 253–277). London: Academic Press.

Melott, A. L. (2008). Long-term cycles in the history of life: Periodic biodiversity in the paleobiology database. *PLoS ONE*, **3**(12) e4044.

Melott, A. L. & R. K. Bambach (2010). Nemesis reconsidered. *Monthly Notices of the Royal Astronomical Society: Letters*, **407**(1) L99–L102.

Melott, A. L. & R. K. Bambach (2011a). A ubiquitous ∼62-Myr periodic fluctuation superimposed on general trends in fossil biodiversity. I. Documentation. *Paleobiology*, **37**(1) 92–112.

Melott, A. L. & R. K. Bambach (2011b). A ubiquitous ∼62-Myr periodic fluctuation superimposed on general trends in fossil biodiversity. II. Evolutionary dynamics associated with periodic fluctuation in marine diversity. *Paleobiology*, **37**(3) 383–408.

Melott, A. L. & R. K. Bambach (2013). Do periodicities in extinction – with possible astronomical connections – survive a revision of the geological timescale? *The Astrophysical Journal*, **773**(1) 6.

Melott, A. L. & R. K. Bambach (2014). Analysis of periodicity of extinction using the 2012 geological time scale. *Paleobiology*, **40**(2) 177–196.

Muller, R. A., P. Hut, M. Davis & W. Alvarez (1984). Cometary showers and unseen solar companions (reply). *Nature*, **312**(5992) 380–381.

Nater, A., M. P. Mattle-Greminger, A. Nurcahyo, et al. (2017). Morphometric, behavioral, and genomic evidence for a new orangutan species. *Current Biology*, **27**(22) 3487–3498.

Newell, N. D. (1952). Periodicity in invertebrate evolution. *Journal of Paleontology*, **26**(3) 371–385.

Newell, N. D. (1962). Paleontological gaps and geochronology. *Journal of Paleontology*, **36**(3) 592–610.

Newell, N. D. (1965). Mass extinctions at the end of the Cretaceous Period. *Science*, **149**(3687) 922–924.

Newell, N. D. (1967). Revolutions in the history of life. *Geological Society of America Special Papers*, **89** 63–92.

Rampino, M. R. (2015). Disc dark matter in the Galaxy and potential cycles of extraterrestrial impacts, mass extinctions and geological events. *Monthly Notices of the Royal Astronomical Society*, **448**(2) 1816–1820.

Rampino, M. R. & K. Caldeira (2015). Periodic impact cratering and extinction events over the last 260 million years. *Monthly Notices of the Royal Astronomical Society*, **454**(4) 3480–3484.

Rampino, M. R. & R. B. Stothers (1984). Terrestrial mass extinctions, cometary impacts and the Sun's motion perpendicular to the Galactic Plane. *Nature*, **308**, 709.

Raup, D. M. (1992). Large-body impact and extinction in the Phanerozoic. *Paleobiology*, **18**(1) 80–88.

Raup, D. M., & J. J. Sepkoski (1984). Periodicity of extinctions in the geologic past. *Proceedings of the National Academy of Sciences*, **81**(3) 801–805. doi:10.1073/pnas.81.3.801.

Raup, D. M. & J. J. Sepkoski (1986). Periodic extinction of families and genera. *Science*, **231**(4740) 833–836.

Raup, D. M. & J. J. Sepkoski (1988). Testing for periodicity of extinction. *Science*, **241**(4861) 94–96.

Rohde, R. A. & R. A. Muller (2005). Cycles in fossil diversity. *Nature*, **434** 208–210.

Schulte, P., L. Alegret, I. Arenillas, et al. (2010). The Chicxulub asteroid impact and mass extinction at the Cretaceous–Paleogene boundary. *Science*, **327**(5970) 1214–1218.

Schwartz, R. D. & P. B. James (1984). Periodic mass extinctions and the Sun's oscillation about the Galactic Plane. *Nature*, **308**(5961) 709–712.

Tabor, C. R., C. G. Bardeen, B. L. Otto-Bliesner, R. R. Garcia & O. B. Toon (2020). Causes and climatic consequences of the impact winter at the Cretaceous–Paleogene boundary. *Geophysical Research Letters*, **47**(3) https://doi.org/10.1029/2019GL085572

Weissman, P. R. (1984). Cometary showers and unseen solar companions. *Nature*, **312** (5992) 380–381.

Whitmire, D. P. & A. A. Jackson (1984). Are periodic mass extinctions driven by a distant solar companion? *Nature*, **308** (5992) 713–715.

Wolfe, J. A. (1991). Palaeobotanical evidence for a June impact winter at the Cretaceous/Tertiary boundary. *Nature*, **352**(6334) 420–423.

CAUSES OF THE END-PERMIAN AND END-CRETACEOUS EXTINCTION EVENTS

Abrajevitch, A., E. Font, F. Florindo & A. P. Roberts (2015). Asteroid impact vs. Deccan eruptions: The origin of low magnetic susceptibility beds below the Cretaceous–Paleogene boundary revisited. *Earth and Planetary Science Letters*, **430** 209–223.

Alroy, J. (2008). Dynamics of origination and extinction in the marine fossil record. *Proceedings of the National Academy of Sciences USA*, **105** (Supplement 1) 11536–11542.

Alroy, J., M. Aberhan, D. J. Bottjer, et al. (2008). Phanerozoic trends in the global diversity of marine invertebrates. *Science*, **321**(5885) 97–100.

Alvarez, L. W., W. Alvarez, F. Asaro & H. V. Michel (1980). Extraterrestrial cause for the Cretaceous–Tertiary extinction. *Science*, **208**(4448) 1095–1108.

Alvarez, W. (2003). Comparing the evidence relevant to impact and flood basalt at times of major mass extinctions. *Astrobiology*, 3(1) 153–161.

Archibald, J., W. A. Clemens, K. Padian, et al. (2010). Cretaceous extinctions: Multiple causes. *Science*, **328**(5981) 973.

Arens, N. C. & I. D. West (2008). Press-pulse: A general theory of mass extinction? *Paleobiology*, **34**(4) 456–471.

Artemieva, N. & J. Morgan (2020). Global K–Pg layer deposited from a dust cloud. *Geophysical Research Letters*, **47**(6) https://doi.org/10.1029/2019gl086562

Baresel, B., H. Bucher, B. Bagherpour, et al. (2017). Timing of global regression and microbial bloom linked with the Permian–Triassic boundary mass extinction: Implications for driving mechanisms. *Scientific Reports*, **7**(1) 43630.

Berner, R. A. & D. J. Beerling (2007). Volcanic degassing necessary to produce a $CaCO_3$ undersaturated ocean at the Triassic–Jurassic boundary. *Palaeogeography, Palaeoclimatology, Palaeoecology*, **244**(1–4) 368–373.

Black, B. A., J.-F. Lamarque, C. A. Shields, L. T. Elkins-Tanton & J. T. Kiehl (2013). Acid rain and ozone depletion from pulsed Siberian Traps magmatism. *Geology*, **42**(1) 67–70.

Bonis, N. R. & W. M. Kürschner (2012). Vegetation history, diversity patterns, and climate change across the Triassic/Jurassic boundary. *Paleobiology*, **38**(2) 240–264.

Bramlette, M. N. (1965). Massive extinctions in biota at the end of Mesozoic time. *Science*, **148**(3678) 1696–1699.

Broadley, M. W., P. H. Barry, C. J. Ballentine, L. A. Taylor & R. Burgess (2018). End-Permian extinction amplified by plume-induced release of recycled lithospheric volatiles. *Nature Geoscience*, **11**(9) 682–687.

Burgess, S. D., S. A. Bowring & S. Shen (2014). High-precision timeline for Earth's most severe extinction. *Proceedings of the National Academy of Sciences USA*, **111**(9) 3316–3321.

Capriolo, M., A. Marzoli, L. E. Aradi, et al. (2020). Deep CO_2 in the end-Triassic Central Atlantic magmatic province. *Nature Communications*, **11**(1) https://doi.org/10.1038/s41467-020-15325-6

Chen, Z.-Q. & M. J. Benton (2012). The timing and pattern of biotic recovery following the end-Permian mass extinction. *Nature Geoscience*, **5**(6) 375–383.

Chenet, A.-L., V. Courtillot, F. Fluteau, et al. (2009). Determination of rapid Deccan eruptions across the Cretaceous–Tertiary boundary using paleomagnetic secular variation: 2. Constraints from analysis of eight new sections and synthesis for a 3500-m-thick composite section. *Journal of Geophysical Research*, **114**(B6) B06103.

Clapham, M. E. & P. R. Renne (2019). Flood basalts and mass extinctions. *Annual Review of Earth and Planetary Sciences*, **47**(1) 275–303.

Courtillot, V. & F. Fluteau (2010). Cretaceous extinctions: The volcanic hypothesis. *Science*, **328**(5981) 973–974.

DePalma, R. A., J. Smit, D. A. Burnham, et al. (2019). A seismically induced onshore surge deposit at the KPg boundary, North Dakota. *Proceedings of the National Academy of Sciences USA*, **116**(17) 8190–8199.

Editorial (2013). The upside of impacts. *Nature Geoscience*, **6**(12) 987.

Erwin, D. H. (2001). Lessons from the past: Biotic recoveries from mass extinctions *Proceedings of the National Academy of Sciences USA*, **98**(10) 5399–5403.

Erwin, D. H. (2006). *Extinction. How Life Nearly Died 250 Million Years Ago.* Princeton, New Jersey: Princeton University Press. (Book)

Fields, B. D., A. L. Melott, J. Ellis, et al. (2020). Supernova triggers for end-Devonian extinctions. *Proceedings of the National Academy of Sciences USA*, **117**(35) 21008–21010.

Georgiev, S., H. Stein, J. Hannah & B. Bingen (2011). Hot acidic Late Permian seas stifle life in record time. *Earth and Planetary Science Letters*, **310**(3–4) 389–400.

Gertsch, B., G. Kelly, T. Adarry & R. Garg (2011). Environmental effects of Deccan volcanism across the Cretaceous–Tertiary transition in Meghalaya, India. *Earth and Planetary Science Letters*, **310**(3–4) 272–285.

Gulick, S. P. S., T. J. Bralower, J. Ormö, et al. (2019). The first day of the Cenozoic. *Proceedings of the National Academy of Sciences USA*, **116**(39) 19342–19351.

Hull, P. M., A. Bornemann, D. E. Penman, et al. (2020). On impact and volcanism across the Cretaceous–Paleogene boundary. *Science*, **367**(6475) 266–272.

Jones, D. S., A. Martini, D. Fike & K. Kaiho (2017). A volcanic trigger for the Late Ordovician mass extinction? Mercury data from south China and Laurentia. *Geology*, **45**(7) 631–634.

Keller, G. (1993). The Cretaceous–Tertiary boundary transition in the Antarctic Ocean and its global implications. *Marine Micropaleontology*, **21**(1–3) 1–45.

Keller, G., T. Adatte, W. Stinnesbeck, et al. (1997). The Cretaceous–Tertiary transition on the shallow Saharan Platform of southern Tunisia. *Geobios*, **30**(7) 951–975.

Keller, G., H. Armstrong, V. Courtillot, et al. (2012). Volcanism, impacts and mass extinctions (long version). *Geoscientist Online* http://www.geolsoc.org .uk/Geoscientist/Archive/November-2012/Volcanism-impacts-and-mass-extinctions-2

Knoll, A. H., R. K. Bambach, J. L. Payne, S. Pruss & W. W. Fischer (2007). Paleophysiology and end-Permian mass extinction. *Earth and Planetary Science Letters*, **256**(3) 295–313.

Kornei, K. (2018). Huge global tsunami followed dinosaur-killing asteroid impact. *EOS*, **99** https://doi.org/10.1029/2018EO112419

Lindström, S., H. Sanei, B. van de Schootbrugge, et al. (2019). Volcanic mercury and mutagenesis in land plants during the end-Triassic mass extinction. *Science Advances*, **5**(10) eaaw4018.

Linzmeier, B. J., A. D. Jacobson, B. B. Sageman, et al. (2019). Calcium isotope evidence for environmental variability before and across the Cretaceous–Paleogene mass extinction. *Geology*. **48**(48) 34–38.

Looy, C. V., R. J. Twitchett, D. L. Dilcher, et al. (2001). Life in the end-Permian dead zone. *Proceedings of the National Academy of Sciences USA*, **98**(14) 7879–7883.

MacLeod, N. (2003). Causes of Phanerozoic extinctions. In L. Rothschild & A. Lister (Eds.), *Evolution on Planet Earth* (pp. 253–277). London: Academic Press.

Paul, C. R. C. (2005). Interpreting bioevents: What exactly did happen to planktonic foraminifers across the Cretaceous–Tertiary boundary? *Palaeogeography, Palaeoclimatology, Palaeoecology*, **224**(1) 291–310.

Petersen, S. V., A. Dutton & K. C. Lohmann (2016). End-Cretaceous extinction in Antarctica linked to both Deccan volcanism and meteorite impact via climate change. *Nature Communications*, **7**(1) 12079.

Phipps, M. J., T. J. Reston & C. R. Ranero (2004). Contemporaneous mass extinctions, continental flood basalts, and 'impact signals': Are mantle plume-induced lithospheric gas explosions the causal link? *Earth and Planetary Science Letters*, **217**(3) 263–284.

Prauss, M. L. (2009). The K/Pg boundary at Brazos-River, Texas, USA – an approach by marine palynology. *Palaeogeography, Palaeoclimatology, Palaeoecology*, **283**(3–4) 195–215.

Retallack, G. J., J. J. Veevers & R. Morante (1996). Global coal gap between Permian–Triassic extinction and Middle Triassic recovery of peat-forming plants. *Geological Society of America Bulletin*, **108** 195–207.

Richoz, S., B. van de Schootbrugge, J. Pross, et al. (2012). Hydrogen sulphide poisoning of shallow seas following the end-Triassic extinction. *Nature Geoscience*, 5(9) 662–667.

Robertson, D. S., W. M. Lewis, P. M. Sheehan & O. B. Toon (2013). K–Pg extinction: Re-evaluation of the heat-fire hypothesis. *Journal of Geophysical Research: Biogeosciences*, 118(1) 329–336.

Rothman, D. H., G. P. Fournier, K. L. French, et al. (2014). Methanogenic burst in the end-Permian carbon cycle. *Proceedings of the National Academy of Sciences USA*. https://doi.org/10.1073/pnas.1318106111

Schulte, P., L. Alegret, I. Arenillas, et al. (2010). The Chicxulub asteroid impact and mass extinction at the Cretaceous–Paleogene boundary. *Science*, 327(5970) 1214–1218.

Song, H., P. B. Wignall, D. Chu, et al. (2014). Anoxia/high temperature double whammy during the Permian–Triassic marine crisis and its aftermath. *Scientific Reports*, 4 4132.

Stordal, F., H. H. Svensen, I. Aarnes & M. Roscherd (2017). Global temperature response to century-scale degassing from the Siberian Traps large igneous province. *Palaeogeography, Palaeoclimatology, Palaeoecology*, 471 96–107.

Tabor, C. R., C. G. Bardeen, B. L. Otto-Bliesner, R. R. Garcia & O. B. Toon (2020). Causes and climatic consequences of the impact winter at the Cretaceous–Paleogene boundary. *Geophysical Research Letters*, 47(3) https://doi.org/10.1029/2019gl085572

Tarailo, D. A. & D. E. Fastovsky (2012). Post-Permo-Triassic terrestrial vertebrate recovery: Southwestern United States. *Paleobiology*, 38(4) 644–663.

Tyrrell, T., A. Merico & D. I. Armstrong McKay. (2015). Severity of ocean acidification following the end-Cretaceous asteroid impact. *Proceedings of the National Academy of Sciences USA*, 112(21) 6556–6561.

van de Schootbrugge, B., T. M. Quan, S. Lindström, et al. (2009). Floral changes across the Triassic/Jurassic boundary linked to flood basalt volcanism. *Nature Geoscience* 2(8) 589–594.

Visscher, H., M. A. Sephton & C. V. Looy (2011). Fungal virulence at the time of the end-Permian biosphere crisis? *Geology*, 39 883–886.

Ward, P. D., J. W. Haggart, E. S. Carter, et al. (2001). Sudden productivity collapse associated with the Triassic–Jurassic boundary mass extinction. *Science*, 292 (5519) 1148–1151.

White, R. V. & A. D. Saunders (2005). Volcanism, impact and mass extinctions: Incredible or credible coincidences? *Lithos*, 79(3) 299–316.

Wignall, P. B. (2001). Large igneous provinces and mass extinctions. *Earth-Science Reviews*, 53(1–2) 1–33.

Wignall, P. B. (2005). The link between large igneous province eruptions and mass extinctions. *Elements*, **1**(5) 293–297.

Wignall, P. B. (2015). *The Worst of Times*. Princeton, New Jersey: Princeton University Press. (Book)

TIME HEALS ALL: RECOVERING FROM A MASS EXTINCTION

Atkinson, J. W. & P. B. Wignall (2019). How quick was marine recovery after the end-Triassic mass extinction and what role did anoxia play? *Palaeogeography, Palaeoclimatology, Palaeoecology*, **528**, 99–119.

Barreda, V. D., N. R. Cúneo, P. Wilf, et al. (2012). Cretaceous/Paleogene floral turnover in Patagonia: Drop in diversity, low extinction, and a Classopollis spike. *PLoS ONE*, **7**(12) e52455.

Henehan, M. J., A. Ridgwell, E. Thomas, et al. (2019). Rapid ocean acidification and protracted Earth system recovery followed the end-Cretaceous Chicxulub impact. *Proceedings of the National Academy of Sciences USA*, 116(45) 22500–22504.

Hull, P. (2015). Life in the aftermath of mass extinctions. *Current Biology*, 25(19) R941–R952.

Lowery, C. M. & A. J. Fraass (2019). Morphospace expansion paces taxonomic diversification after end Cretaceous mass extinction. *Nature Ecology and Evolution*, **3**(6) 900–904.

Lyson, T. R., Miller, I. M., Bercovici, A. D., et al. (2019). Exceptional continental record of biotic recovery after the Cretaceous–Paleogene mass extinction. *Science*, **366**(6468) 977–983.

Rodríguez-Tovar, F. J., C. M. Lowery, T. J. Bralower, S. P. S. Gulick & H. L. Jones (2020). Rapid macrobenthic diversification and stabilization after the end-Cretaceous mass extinction event. *Geology* https://doi.org/10.1130/g47589.1

Vajda, V. & McLoughlin, S. (2007). Extinction and recovery patterns of the vegetation across the Cretaceous–Palaeogene boundary – a tool for unravelling the causes of the end-Permian mass-extinction. *Review of Palaeobotany and Palynology*, **144**(1–2) 99–112.

Vajda, V., Raine, J. I. & Hollis, C. J. (2001). Indication of global deforestation at the Cretaceous–Tertiary boundary by New Zealand fern spike. *Science*, **294**(5547) 1700–1702.

Whittle, R. J., Witts, J. D., Bowman, V. C., et al. (2019). Nature and timing of biotic recovery in Antarctic benthic marine ecosystems following the Cretaceous–Palaeogene mass extinction. *Palaeontology*, **62**(6) 919–934.

THE LATE QUATERNARY MEGAFAUNAL EXTINCTIONS

Alley, R. B. (2004). GISP2 ice core temperature and accumulation data. IGBP PAGES/World Data Center for Paleoclimatology Data Contribution Series #2004-013. Boulder, Colorado: NOAA/NGDC Paleoclimatology Program.

Ardelean, C. F., L. Becerra-Valdivia, M. W. Pedersen et al. (2020). Evidence of human occupation in Mexico around the Last Glacial Maximum. *Nature*, **584** (7819) 87–92.

Barnosky, A. D. (2008). Megafauna biomass tradeoff as a driver of Quaternary and future extinctions. *Proceedings of the National Academy of Sciences USA* **105** (Supplement 1) 11543–11548.

Barnosky, A. D., P. L. Koch, R. S. Feranec, S. L. Wing & A. B. Shabel (2004). Assessing the causes of Late Pleistocene extinctions on the continents. *Science* **306**(5693) 70–75.

Barnosky, A. D. & E. L. Lindsey (2010). Timing of Quaternary megafaunal extinction in South America in relation to human arrival and climate change. *Quaternary International* **217**(1) 10–29.

Barnosky, A. D., E. L. Lindsey, N. A. Villavicencio, et al. (2016). Variable impact of late-Quaternary megafaunal extinction in causing ecological state shifts in North and South America. *Proceedings of the National Academy of Sciences USA* **113**(4) 856–861.

Bartlett, L. J., D. R. Williams, G. W. Prescott, et al. (2016). Robustness despite uncertainty: regional climate data reveal the dominant role of humans in explaining global extinctions of Late Quaternary megafauna. *Ecography*, **39**(2) 152–161.

Barbuzano, J. (2020). Armageddon at 10,000 BCE. *EOS*, **101** https://doi.org/10.1029/2020EO142127

Becerra-Valdivia, L., M. R. Waters, T. W. Stafford Jr, et al. (2018). Reassessing the chronology of the archaeological site of Anzick. *Proceedings of the National Academy of Sciences USA*, **115**(27) 7000–7003.

Bocherens, H. (2018). The rise of the Anthroposphere since 50,000 years: An ecological replacement of megaherbivores by humans in terrestrial ecosystems? *Frontiers in Ecology and Evolution*, **6**(3) https://doi.org/10.3389/fevo.2018.00003

Brook, B. W., D. M. J. S. Bowman, D. A. Burney & T. F. Flannery (2007). Would the Australian megafauna have become extinct if humans had never colonised the continent? Comments on 'A review of the evidence for a human role in the extinction of Australian megafauna and an alternative explanation' by S. Wroe and J. Field. *Quaternary Science Reviews*, **26**(3) 560–564.

Burney, D. A. & T. F. Flannery (2005). Fifty millennia of catastrophic extinctions after human contact. *Trends in Ecology and Evolution*, **20**(7) 395–401.

Carrasco, M. A., A. D. Barnosky & R. W. Graham (2009). Quantifying the extent of North American mammal extinction relative to the pre-Anthropogenic baseline. *PLoS ONE*, **4**(12) e8331.

Ceballos, G. & P. R. Ehrlich (2002). Mammal population losses and the extinction crisis. *Science*, **296**(5569) 904–907.

Charles, R. K., S. S. Que Hee, A. Stich, et al. (2014). Nanodiamond-rich layer across three continents consistent with major cosmic impact at 12,800 Cal BP. *The Journal of Geology*, **122**(5) 475–506.

Cooper, A., C. Turney, K. A. Hughen, et al. (2015). Abrupt warming events drove Late Pleistocene Holarctic megafaunal turnover. *Science*, **349**(6248) 602–606.

Daulton, T. L., S. Amari, A. C. Scott, et al. (2017). Comprehensive analysis of nanodiamond evidence relating to the Younger Dryas impact hypothesis. *Journal of Quaternary Science*, **32**(1) 7–34.

Firestone, R. B., West, A., Kennett, J. P., et al. (2007). Evidence for an extraterrestrial impact 12,900 years ago that contributed to the megafaunal extinctions and the Younger Dryas cooling. *Proceedings of the National Academy of Sciences USA*, **104**(41), 16016–16021

Grayson, D. K. & D. J. Meltzer (2002). Clovis hunting and large mammal extinction: A critical review of the evidence. *Journal of World Prehistory*, **16**(4) 313–359.

Grayson, D. K. & D. J. Meltzer (2003). A requiem for North American overkill. *Journal of Archaeological Science*, **30**(5) 585–593.

Halligan, J. J., Waters, M. R., Perrotti, A., et al. (2016). Pre-Clovis occupation 14,550 years ago at the Page-Ladson site, Florida, and the peopling of the Americas. *Science Advances*, **2**(5) e1600375.

Holdaway, R. N. & C. Jacomb (2000). Rapid extinction of the moas (*Aves dinornithiformes*): Model, test, and implications. *Science*, **287**(5461) 2250–2254.

Johnson, C. N., C. J. A. Bradshaw, A. Cooper, R. Gillespie & B. W. Brook (2013). Rapid megafaunal extinction following human arrival throughout the New World. *Quaternary International*, **308–309** 273–277.

Kinzie, C. R., S. S. Que Hee, A. Stich, et al. (2014). Nanodiamond-rich layer across three continents consistent with major cosmic impact at 12,800 Cal BP. *The Journal of Geology*, **122**(5) 475–506.

Kjær, K. H., N. K. Larsen, T. Binder, et al. (2018). A large impact crater beneath Hiawatha Glacier in northwest Greenland. *Science Advances*, **4**(11) eaar8173.

Koch, P. L. & A. D. Barnosky (2006). Late Quaternary extinctions: State of the debate. *Annual Review of Ecology, Evolution, and Systematics*, **37**(1) 215–250.

Lima-Ribeiro, M. S. & J. A. Felizola Diniz-Filho (2013). American megafaunal extinctions and human arrival: Improved evaluation using a meta-analytical approach. *Quaternary International*, **299** 38–52.

Mann, D. H., P. Groves, B. V. Gaglioti & B. A. Shapiro (2019). Climate-driven ecological stability as a globally shared cause of Late Quaternary megafaunal extinctions: The Plaids and Stripes Hypothesis. *Biological Reviews*, **94**(1) 328–352.

Mann, D. H., P. Groves, R. E. Reanier, et al. (2015). Life and extinction of megafauna in the ice-age Arctic. *Proceedings of the National Academy of Sciences USA*, **112**(46) 14301–14306.

Martin, P. S. (1966). Africa and Pleistocene overkill. *Nature*, **212** (5075) 1615–1616.

Martin, P. S. (1990). 40,000 years of extinctions on the 'planet of doom'. *Palaeogeography, Palaeoclimatology, Palaeoecology*, **82**(1) 87–201.

Mauro, G., M. Moleón, P. Jordano, et al. (2018). Ecological and evolutionary legacy of megafauna extinctions. *Biological Reviews*, **93**(2) 845–862.

Meltzer, D. J., V. T. Holliday, M. D. Cannon & D. S. Miller (2014). Chronological evidence fails to support claim of an isochronous widespread layer of cosmic impact indicators dated to 12,800 years ago. *Proceedings of the National Academy of Sciences USA*, **111**(21) E2162–E2171.

Miller, G. H., M. L. Fogel, J. W. Magee, et al. (2005). Ecosystem collapse in Pleistocene Australia and a human role in megafaunal extinction. *Science*, **309** (5732) 287–290.

Moore, A. M. T., Kennett, J. P., Napier, W. M., et al. (2020). Evidence of cosmic impact at Abu Hureyra, Syria at the Younger Dryas onset (~12.8 ka): High-temperature melting at >2200 °C. *Scientific Reports*, 10(1) 4185.

O'Connell, J. F., J. Allen, M. A. J. Williams, et al. (2018). When did *Homo sapiens* first reach Southeast Asia and Sahul? *Proceedings of the National Academy of Sciences USA*, **115**(34) 8482–8490.

Pääbo, S. (2014) *Neanderthal Man, In Search of Lost Genomes*. New York: Basic Book (Book)

Perry, G. L. W., A. B. Wheeler, J. R. Wood & J. M. Wilmshurst (2014). A high-precision chronology for the rapid extinction of New Zealand moa (Aves, Dinornithiformes). *Quaternary Science Reviews*, **105** 126–135.

Pimiento, C., J. Griffin, C. F. Clements, et al. (2017). The Pliocene marine megafauna extinction and its impact on functional diversity. *Nature Ecology and Evolution*, **1**(8) 1100–1106.

Pinter, N., A. C. Scott, T. L. Daulton, et al. (2011). The Younger Dryas impact hypothesis: A requiem. *Earth-Science Reviews*, 106(3–4) 247–264.

Pino, M., A. M. Abarzúa, G. Astorga, et al. (2019). Sedimentary record from Patagonia, southern Chile supports cosmic-impact triggering of biomass burning, climate change, and megafaunal extinctions at 12.8 ka. *Scientific Reports*, 9(1) 4413.

Politis, G. G., M. A. Gutiérrez, D. J. Rafuse & A. Blasi (2016). The arrival of *Homo sapiens* into the Southern Cone at 14,000 years ago. *PLoS ONE*, **11**(9) e0162870.

Prescott, G. W., D. R. Williams, A. Balmford, R. E. Green & A. Manica (2012). Quantitative global analysis of the role of climate and people in explaining late Quaternary megafaunal extinctions. *Proceedings of the National Academy of Sciences USA*, **109**(12) 4527–4531.

Prideaux, G. J., G. Gully, A. Couzens, et al. (2010). Timing and dynamics of Late Pleistocene mammal extinctions in southwestern Australia. *Proceedings of the National Academy of Sciences USA*, **107**(51) 22157–22162.

Reich, D. (2018) *Who We Are and How We Got Here*, Oxford: Oxford University Press (Book)

Rick, T. C., P. V. Kirch, J. M. Erlandson & S. M. Fitzpatricke (2013). Archeology, deep history, and the human transformation of island ecosystems. *Anthropocene*, **4** 33–45.

Robinson, G. S., L. Pigott Burney & D. A. Burney (2005). Landscape paleoecology and megafaunal extinction in southeastern New York State. *Ecological Monographs*, **75**(3) 295–315.

Rule, S., B. W. Brook, S. G. Haberle, et al. (2012). The aftermath of megafaunal extinction ecosystem transformation in Pleistocene Australia. *Science*, **335** (6075) 1483–1486.

Saltré, F., M. Rodríguez-Rey, B. W. Brook, et al. (2016). Climate change not to blame for late Quaternary megafauna extinctions in Australia. *Nature Communications*, **7** 10511.

Sandom, C., S. Faurby, B. Sandel & J.-C. Svenning (2014). Global late Quaternary megafauna extinctions linked to humans, not climate change. *Proceedings of the Royal Society B: Biological Sciences*, **281**(1787) https://doi.org/10.1098/rspb.2013.3254

Seersholm, F. V., D. J. Werndly, A. Grealy, et al. (2020). Rapid range shifts and megafaunal extinctions associated with late Pleistocene climate change. *Nature Communications*, **11**(1) https://doi.org/10.1038/s41467-020-16502-3

Shillito, L.-M., H. L. Whelton, J. C. Blong, et al. (2020). Pre-Clovis occupation of the Americas identified by human faecal biomarkers in coprolites from Paisley Caves, Oregon. *Science Advances*, **6**(29) eaba6404.

Steadman, D. W. (1995). Prehistoric extinctions of Pacific Island birds: Biodiversity meets zooarchaeology. *Science*, **267**(5201) 1123–1131.

Tobler, R., A. Rohrlach, J. Soubrier, et al. (2017). Aboriginal mitogenomes reveal 50,000 years of regionalism in Australia. *Nature*, **544** 180–184.

van der Kaars, S., G. H. Miller, C. S. M. Turney, et al. (2017). Humans rather than climate the primary cause of Pleistocene megafaunal extinction in Australia. *Nature Communications*, **8** 14142.

Villavicencio, N. A., E. L. Lindsey, F. M. Martin, et al. (2016). Combination of humans, climate, and vegetation change triggered Late Quaternary megafauna extinction in the Última Esperanza region, southern Patagonia, Chile. *Ecography*, **39**(2) 125–140.

Waters, M. R., J. L. Keene, S. L. Forman, et al. (2018). Pre-Clovis projectile points at the Debra L. Friedkin site, Texas – Implications for the Late Pleistocene peopling of the Americas. *Science Advances*, **2**(5) eaat4505.

Westaway, M. C., J. Olley & R. Grün (2017). At least 17,000 years of coexistence: Modern humans and megafauna at the Willandra Lakes, South-Eastern Australia. *Quaternary Science Reviews*, **157** 206–211.

Wood, J. R., J. A. Alcover, T. M. Blackburn, et al. (2017). Island extinctions: Processes, patterns, and potential for ecosystem restoration. *Environmental Conservation*, **44**(4) 348–358.

Wroe, S. & J. Field (2006). A review of the evidence for a human role in the extinction of Australian megafauna and an alternative interpretation. *Quaternary Science Reviews*, **25**(21) 2692–2703.

Wroe, S., J. Field, R. Fullagar & L. S. Jermin (2004). Megafaunal extinction in the late Quaternary and the global overkill hypothesis. *Alcheringa: An Australasian Journal of Palaeontology*, **28**(1) 291–331.

Wroe, S., J. Field & D. K. Grayson (2006). Megafaunal extinction: Climate, humans and assumptions. *Trends in Ecology and Evolution*, **21**(2) 61–62.

Zeberg, H. & S. Pääbo. (2020). The major genetic risk factor for severe COVID-19 is inherited from Neanderthals. *Nature*, **587** 610–612.

Zelenkov, N. V., Lavrov, A. V., Startsev, D. B., Vislobokova, I. A. & Lopatin, A. V. (2019). A giant early Pleistocene bird from eastern Europe: Unexpected component of terrestrial faunas at the time of early Homo arrival. *Journal of Vertebrate Paleontology*, **39**(2) e1605521

SURVIVING THE ANTHROPOCENE

Barnosky, A. D., E. A. Hadly, J. Bascompte, et al. (2012). Approaching a state shift in Earth's biosphere. *Nature*, **486**(7401) 52–58.

Barnosky, A. D., N. Matzke, S. Tomiya, et al. (2011). Has the Earth's sixth mass extinction already arrived? *Nature*, **471**(7336) 51–57.

Berentson, Q. (2012). *Moa*. Nelson: Craig Potten Publishing. (Book)

Braje, T. J. (2015). Earth Systems, human agency, and the Anthropocene: Planet Earth in the Human Age. *Journal of Archaeological Research*, **23**(4) 369–396.

Brook, B. W., N. S. Sodhi & C. J. A. Bradshaw (2008). Synergies among extinction drivers under global change. *Trends in Ecology and Evolution*, **23**(8) 453–460.

Budiansky, S. (1994). Extinction or miscalculation? *Nature* **370**(6485) 104.

Ceballos, G. & P. R. Ehrlich (2002). Mammal population losses and the extinction crisis. *Science*, **296**(5569) 904–907.

Ceballos, G. & P. R. Ehrlich (2018). The misunderstood sixth mass extinction. *Science*, **360**(6393) 1080–1081.

Ceballos, G., P. R. Ehrlich, A. D. Barnosky, et al. (2015). Accelerated modern human-induced species losses: Entering the sixth mass extinction. *Science Advances*, **1**(5) e1400253.

Ceballos, G., P. R. Ehrlich & R. Dirzo (2017). Biological annihilation via the ongoing sixth mass extinction signalled by vertebrate population losses and declines. *Proceedings of the National Academy of Sciences USA*, **114**(30) E6089–E6096.

Ceballos, G., A. Garcia & P. R. Ehrlich (2010). The Sixth Extinction Crisis loss of animal populations and species. *Journal of Cosmology*, **8** 1821–1831.

Davis, M., S. Faurby & J.-C. Svenning (2018). Mammal diversity will take millions of years to recover from the current biodiversity crisis. *Proceedings of the National Academy of Sciences USA*, **115**(44) 11262–11267.

Dirzo, R. & P. H. Raven (2003). Global state of biodiversity and loss. *Annual Review of Environment and Resources*, **28**(1) 137–167.

Dirzo, R., H. S. Young, M. Galetti, et al. (2014). Defaunation in the Anthropocene. *Science*, **345**(6195) 401–406.

Duffy, J. E. (2009). Why biodiversity is important to the functioning of real-world ecosystems. *Frontiers in Ecology and the Environment*, **7**(8) 437–444.

Frieling, J., H. H. Svensen, S. Planke, et al. (2016). Thermogenic methane release as a cause for the long duration of the PETM. *Proceedings of the National Academy of Sciences USA*, **113**(43) 12059–12064.

Gutjahr, M., A. Ridgwell, P. F. Sexton, et al. (2017). Very large release of mostly volcanic carbon during the Palaeocene–Eocene Thermal Maximum. *Nature*, 548(7669) 573–577.

IUCN (2020). The IUCN Red List of Threatened Species. Version 2020-2. https://www.iucnredlist.org (Accessed 18 September 2020.)

McInerney, F. A. & S. L. Wing (2011). The Paleocene–Eocene Thermal Maximum: A perturbation of carbon cycle, climate, and biosphere with implication for the future. *Annual Review of Earth and Planetary Science*, **39** 489–516.

Myers, N., R. A. Mittermeier, C. G. Mittermeier, G. A. B. da Fonseca & J. Kent (2000). Biodiversity hotspots for conservation priorities. *Nature,* **403**(6772) 853–858.

Pimm, S. L., C. N. Jenkins, R. Abell, et al. (2014). The biodiversity of species and their rates of extinction, distribution, and protection. *Science,* **344**(6187) 1246752.

Pimm, S. L. & P. Raven (2000). Extinction by numbers. *Nature,* **403**(6773) 843–845.

Pimm, S. L., G. J. Russell, J. L. Gittleman & T. M. Brooks (1995). The future of biodiversity. *Science* **269**(5222) 347–350.

Rollinson, H. (2007). *Early Earth Systems.* Malden, MA: Blackwell Publishing. (Book)

Steffen, W., J. Rockström, K. Richardson, et al. (2018). Trajectories of the Earth System in the Anthropocene. *Proceedings of the National Academy of Sciences USA,* **115**(33) 8252–8259.

Strassburg, B. B. N., Iribarrem, A., Beyer, H. L., et al. (2020). Global priority areas for ecosystem restoration. *Nature* https://doi.org/10.1038/s41586-020-2784-9

Tyrrell, T. (2013). *On Gaia.* Princeton, New Jersey: Princeton University Press. (Book)

Zachos, J. C., Dickens, G. R. & Zeebe, R. E. (2008). An early Cenozoic perspective on greenhouse warming and carbon-cycle dynamics. *Nature,* **451**(7176) 279–283.

Index